INTERNATIONAL CENTRE FOR MECHANICAL SCIENCES

COURSES AND LECTURES - No. 32

GIUSEPPE LONGO
UNIVERSITY OF TRIESTE

SOURCE CODING THEORY

LECTURES HELD AT THE DEPARTMENT
FOR AUTOMATION AND INFORMATION
JUNE 1970

UDINE 1970

SPRINGER-VERLAG WIEN GMBH

Originally published by Springer-Verlag Wien-New York 1972

ISBN 978-3-211-81090-3 ISBN 978-3-7091-2842-8 (eBook)
DOI 10.1007/978-3-7091-2842-8

PREFACE

Most of the material contained in these lecture notes was covered by the author during a course held at the International Centre for Mechanical Sciences Udine, in June - July I970.

The topic is coding for discrete memoryless sources without any distortion measure. The results are partly new, at least as regards the approach through the I - divergence concept.

Previous knonwledge of Information Theory or of Statistics is not required, although some familiarity with those fields would help the reader in appreciating how self-contained the notes are, especially as a consequence of the concise elegance of the Preliminaries for which I am indebted to Prof. I. Csiszar.

I am grateful to all Authorities of CISM for giving me the opportunity of delivering the course, and especially to Prof. L. Sobrero whose astonishingly energetic activity is keeping the Centre at an unrivalled standard.

Udine, June 1970

Preliminaries

In this section we summarize some basic definitions and relations which will be used freely in the sequel : the simple proofs will be sketched only.

The term "random variable" will be abbreviated as RV ; for the sake of simplicity, attention will be restricted to the case of discrete RV's, i.e., to RV's with values in a finite or countably infinite set.

ξ, η, ζ will denote RV's with values in the (finite or countably infinite) sets X, Y, Z.

All random variables considered at the same time will be assumed to be defined on the same probability space. Recall that a probability space is a triplet $(\Omega, \mathcal{F}, P)$ where Ω is a set (the set of all conceivable outcomes of an experiment), $\mathcal{F}$ is a σ-algebra of subsets of Ω (the class of observable events) and P is a measure (non-negative countably additive set function) defined on $\mathcal{F}$ such that $P(\Omega)=1$. RV's ξ, η etc. are functions $\xi(\omega)$, $\eta(\omega)$ etc. $(\omega \in \Omega)$. The probability $P\{\xi = x\}$ is the measure of the set of those ω's for which $\xi(\omega)=x$; similarly, $P\{\xi=x, \eta=y\}$ is the measure of the set of those ω's for which $\xi(\omega)=x$ and $\eta(\omega)=y$.

The conditional probability $P\{\xi = x \mid \eta = y\}$ is defined as $\frac{P\{\xi = x, \eta = y\}}{P\{\eta = y\}}$ (if $P\{\eta = y\} = 0$, $P\{\xi = x \mid \eta = y\}$ is undefined).

Definition 1. The RV's defined by

$$(1) \quad \iota_\xi = -\log_2 P\{\xi = x\} \qquad \text{if} \quad \xi = x$$

$$(2) \quad \iota_{\xi \wedge \eta} = \log_2 \frac{P\{\xi = x, \eta = y\}}{P\{\xi = x\} P\{\eta = y\}} \qquad \text{if} \quad \xi = x, \eta = y$$

are called the entropy density of ξ and the information density of ξ and η, respectively.

$$(3) \quad \iota_{\xi \mid \eta} = -\log_2 P\{\xi = x \mid \eta = y\} \qquad \text{if} \quad \xi = x, \eta = y$$

$$(4) \quad \iota_{\xi \mid \eta, \zeta} = -\log_2 P\{\xi = x \mid \eta = y, \zeta = z\} \quad \text{if} \quad \xi = x, \eta = y, \zeta = z$$

are conditional entropy densities and

$$(5) \quad \iota_{\xi \wedge \eta \mid \zeta} = \log_2 \frac{P\{\xi = x, \eta = y \mid \zeta = z\}}{P\{\xi = x \mid \zeta = z\} P\{\eta = y \mid \zeta = z\}} \qquad \text{if} \quad \xi = x, \eta = y, \zeta = z$$

is a conditional information density.

Remark. Entropy density is often called "self-information" and information density "mutual information". In our terminology, the latter term will mean the expectation of $\iota_{\xi \wedge \eta}$.

Definition 2. The quantities

$$E(\xi) \overset{\text{def}}{=} E\iota_\xi = -\sum_{x\in X} P\{\xi=x\} \log_2 P\{\xi=x\} \tag{6}$$

$$I(\xi\wedge\eta) \overset{\text{def}}{=} E\iota_{\xi\wedge\eta} = \sum_{x\in X, y\in Y} P\{\xi=x,\eta=y\} \log_2 \frac{P\{\xi=x,\eta=y\}}{P\{\xi=x\}P\{\eta=y\}} \tag{7}$$

are called the entropy of ξ and the mutual information of ξ and η, respectively.

The quantities

$$H(\xi|\eta) \overset{\text{def}}{=} E\iota_{\xi|\eta} = -\sum_{x\in X, y\in Y} P\{\xi=x,\eta=y\} \log_2 P\{\xi=x|\eta=y\} \tag{8}$$

$$H(\xi|\eta,\zeta) \overset{\text{def}}{=} E\iota_{\xi|\eta,\zeta} = -\sum_{x\in X, y\in Y, z\in Z} P\{\xi=x,\eta=y,\zeta=z\} \log_2 P\{\xi=x|\eta=y,\zeta=z\} \tag{9}$$

are called conditional entropies and

$$I(\xi\wedge\eta|\zeta) \overset{\text{def}}{=} E\iota_{\xi\wedge\eta|\zeta} = \sum_{x\in X, y\in Y, z\in Z} P\{\xi=x,\eta=y,\zeta=z\} \log_2 \frac{P\{\xi=x,\eta=y|\zeta=z\}}{P\{\xi=x|\zeta=z\}P\{\eta=y|\zeta=z\}} \tag{10}$$

is called conditional mutual information.

Here terms like $0 \log_2 0$ or $0 \log_2 \frac{0}{0}$ are meant to be 0.

The quantities (6)-(10) are always non-negative (for (7) and (10) this requires proof ; see (17), (18)) but they may be infinite. The latter contingency should be kept in mind ; in particular, identities like $I(\xi\wedge\eta) = H(\xi) - H(\xi|\eta)$ (cf. (21)) are valid only under the condition that they do not contain the undefined expression $+\infty - \infty$.

$H(\xi)$ is interpreted as the measure of the average amount of information contained in specifying a particular value of ξ; $I(\xi\wedge\eta)$ is a measure of the average amount of information obtained with respect to the value of η when specifying a particular value of ξ. Conditional entropy and conditional mutual information are interpreted similarly. Logarithms to the basis 2 (rather than natural logarithms) are used to ensure that the amount of information provided by a binary digit (more exactly, by a random variable taking on the values 0 and 1 with probabilities 1/2) be unity. This unit of the amount of information is called bit.

The interpretation of the quantities (6)-(10) as measures of the amount of information is not merely a matter of convention ; rather, it is convincingly suggested by a number of theorems of information theory as well as by the great efficency of heuristic reasonings based on this interpretation. There is much less evidence for a similar interpretation of the entropy and information densities. Thus we do not insist on attaching any intuitive meaning to the latters ; they will be used simply as convenient mathematical tools.

A probability distribution, to be abbreviated as PD, on the set X is a non-negative valued function $p(x)$ on X with $\sum_{x\in X} p(x)=1$; PD's will be denoted by script letters, e.g. $\mathcal{P}=\{p(x),\ x\in X\}$.

Definition 3. The I-divergence of two PD's $\mathcal{P} = \{p(x),\ x \in X\}$ and $\mathcal{Q} = \{q(x), x \in X\}$ is defined as

$$I(\mathcal{P} \| \mathcal{Q}) = \sum_{x \in X} p(x) \log_2 \frac{p(x)}{q(x)} \ . \tag{11}$$

Here terms of the form $a \log_2 \frac{a}{0}$ with $a > 0$ are meant to be $+\infty$.

Lemma 1. Using the notations $p(A) = \sum_{x \in A} p(x)$, $q(A) = \sum_{x \in A} q(x)$ we have for an arbitrary subset A of X

$$\sum_{x \in A} p(x) \log_2 \frac{p(x)}{q(x)} \geq p(A) \log_2 \frac{p(A)}{q(A)} \ ; \tag{12}$$

if $q(A) > 0$ the equality holds iff(*) $p(x) = \frac{p(A)}{q(A)} q(x)$ for every $x \in A$. In particular, setting $A = X$:

$$I(\mathcal{P} \| \mathcal{Q}) \geq 0 \ , \qquad \text{equality iff } \mathcal{P} = \mathcal{Q} \ . \tag{13}$$

Proof. The concavity of the function $f(t) = \ln t$ implies $\ln t \leq t - 1$, with equality iff $t = 1$. Setting now $t = \frac{q(x)}{p(x)} \frac{p(A)}{q(A)}$ one gets $\ln \frac{q(x)}{p(x)} \leq \ln \frac{q(A)}{p(A)} + \frac{q(x)}{p(x)} \frac{p(A)}{q(A)} - 1$ whenever $p(x)\, q(x) > 0$, with equality iff $\frac{q(x)}{p(x)} = \frac{q(A)}{p(A)}$.

Multiplying by $p(x)$ and summing for every $x \in A$ with $p(x) > 0$ (one may obviously assume that then $q(x) > 0$ too) (12)follows, including the condition for equality. The choice of the basis of the logarithms being clearly immaterial. The I-divergence $I(\mathcal{P} \| \mathcal{Q})$ is a measure of how different the PD $\mathcal{P}$ is from the PD $\mathcal{Q}$ (however note, that in general $I(\mathcal{P} \| \mathcal{Q}) = I(\mathcal{Q} \| \mathcal{P})$). If $\mathcal{P}$ and $\mathcal{Q}$ are two

(*)Iff is an abbreviation for "if and only if".

hypothetical PD's on X then $I(\mathcal{P}\|\mathcal{Q})$ may be interpreted as the average amount of information in favour of $\mathcal{P}$ and against $\mathcal{Q}$, obtained from observing a randomly chosen element of X, provided that the PD $\mathcal{P}$ is the true one.

The <u>distribution of a RV</u> ξ is the PD $\mathcal{P}_\xi$ defined by

$$(14)\qquad \mathcal{P}_\xi=\{p_\xi(x),\ x\in X\},\quad p_\xi(x)=P\{\xi=x\}.$$

The <u>joint distribution</u> $\mathcal{P}_{\xi\eta}$ of the RV's ξ and η is defined as the distribution of the RV (ξ,η) taking values in $X\times Y$ i.e. $\mathcal{P}_{\xi\eta}=\{p_{\xi\eta}(x,y),\ x\in X, y\in Y\}$, $p_{\xi\eta}(x,y)=P\{\xi=x,\ \eta=y\}$.

From (7) and (11) it follows

$$(15)\qquad I(\xi\wedge\eta)=I(\eta\wedge\xi)=I(\mathcal{P}_{\xi\eta}\|\mathcal{P}_\xi\times\mathcal{P}_\eta)$$

where $\mathcal{P}_\xi\times\mathcal{P}_\eta=\{p_\xi(x)\,p_\eta(y),\ x\in X, y\in Y\}$ and also

$$(16)\qquad I(\xi\wedge\eta)=\sum_{x\in X}p_\xi(x)\,I(\mathcal{P}_{\eta|\xi=x}\|\mathcal{P}_\eta)$$

where $\mathcal{P}_{\eta|\xi=x}=\{p_x(y),\ y\in Y\}$, $p_x(y)=P\{\eta=y\ \ \xi=x\}$.
(15) and (13) yield

(17) $I(\xi\wedge\eta)\geq 0$, equality iff ξ and η are independent.

By a comparison of (7) and (10), this implies

(18) $I(\xi\wedge\eta\mid\zeta)\geq 0$, equality iff ξ and η are condition-

ally independent for ζ given.

Let us agree to write $\iota_{\xi,\eta}$ for $\iota_{(\xi,\eta)}$ (entropy density of the RV (ξ,η)), $\iota_{\xi,\eta\wedge\zeta}$ for $\iota_{(\xi,\eta)\wedge\zeta}$ (information density of the RV's (ξ,η) and ζ) etc. ; omitting the brackets will cause no ambiguities.

Theorem 1. (Basic identities)

$$\iota_{\xi,\eta} = \iota_{\xi|\eta} + \iota_{\eta} \qquad H(\xi,\eta) = H(\xi|\eta) + H(\eta) \tag{19}$$

$$\iota_{\xi,\eta|\zeta} = \iota_{\xi|\eta,\zeta} + \iota_{\eta|\zeta} \qquad H(\xi,\eta|\zeta) = H(\xi|\eta,\zeta) + H(\eta|\zeta) \tag{20}$$

$$\iota_{\xi} = \iota_{\xi|\eta} + \iota_{\xi\wedge\eta} \qquad H(\xi) = H(\xi|\eta) + I(\xi\wedge\eta) \tag{21}$$

$$\iota_{\xi|\zeta} = \iota_{\xi|\eta,\zeta} + \iota_{\xi\wedge\eta|\zeta} \qquad H(\xi|\zeta) = H(\xi|\eta,\zeta) + I(\xi\wedge\eta|\zeta) \tag{22}$$

$$\iota_{\xi_1,\xi_2\wedge\eta} = \iota_{\xi_1\wedge\eta} + \iota_{\xi_2\wedge\eta|\xi_1} \; ; \quad I(\xi_1,\xi_2\wedge\eta) = I(\xi_1\wedge\eta) + I(\xi_2\wedge\eta|\xi_1) \tag{23}$$

$$\iota_{\xi_1,\xi_2\wedge\eta|\zeta} = \iota_{\xi_1\wedge\eta|\zeta} + \iota_{\xi_2\wedge\eta|\xi_1,\zeta} \; ; \quad I(\xi_1,\xi_2\wedge\eta|\zeta) = I(\xi_1\wedge\eta|\zeta) + I(\xi_2\wedge\eta \; \xi_1,\zeta) \tag{24}$$

Proof. Immediate from definitions 1 and 2.

Theorem 2. (Basic inequalities)

The information quantities (6)-(10) are non-negative ;

$$H(\xi,\eta) \geq H(\xi) \, , \quad H(\xi,\eta|\zeta) \geq H(\xi|\zeta) \tag{25}$$

$$H(\xi|\eta,\zeta) \leq H(\xi|\eta) \leq H(\xi) \tag{26}$$

$$I(\xi_1,\xi_2 \wedge \eta) \geqq I(\xi_1 \wedge \eta);\ I(\xi_1,\xi_2 \wedge \eta|\zeta) \geqq$$

(27) $$\geqq I(\xi_1 \wedge \eta|\zeta)$$

(28) $$I(\xi \wedge \eta) \leqq H(\xi),\ I(\xi \wedge \eta|\zeta) \leqq H(\xi|\zeta).$$

If ξ has at most r possible values then

(29) $$H(\xi) \leqq \log_2 r.$$

If ξ has at most $r(y)$ possible values when $\eta = y$ then

(30) $$H(\xi|\eta) \leqq E \log_2 r(\eta).$$

Proof. (25)-(28) are direct consequences of (19)-(24). (29) follows from (13) setting $\mathcal{P} = \mathcal{P}_\xi$, $\mathcal{Q} = \left\{\frac{1}{r}, \ldots, \frac{1}{r}\right\}$; on comparison of (6) and (8), (29) implies (30).

Remark. $I(\xi \wedge \eta|\zeta) \leqq I(\xi \wedge \eta)$ is not valid; in general. E. g., if ξ and η are independent but not conditionally independent for a given ζ, then

$$I(\xi \wedge \eta) = 0 < I(\xi \wedge \eta|\zeta).$$

Theorem 3. (Substitutions in the information quantities).

For arbitrary functions $f(x)$, $f(y)$ or $f(x,y)$ defined on X, Y or $X \times Y$, respectively, the following inequalities hold

(31) $$H(f(\xi)) \leqq H(\xi);\quad I(f(\xi) \wedge \eta) \leqq I(\xi \wedge \eta)$$

$$H(\xi \mid f(\eta)) \geqq H(\xi \mid \eta) \tag{32}$$

$$H(f(\xi,\eta) \mid \eta) \leqq H(\xi \mid \eta). \tag{33}$$

If f is one-to-one, or $f(x,y)$ as a function of x is one-to-one for every fixed $y \in Y$, respectively, the equality signs are valid. In the second half of (31) and in (32) the equality holds also if ξ and η are conditionally independent for given $f(\xi)$ or $f(\eta)$, respectively.

Proof. In the one-to-one case, the validity of (31)-(33) with the equality sign is obvious from definition 2. In the general case, apply this observation for $\tilde{f}$ instead of f where $\tilde{f}(x) = (x, f(x))$, $\tilde{f}(y) = (y, f(y))$ or $\tilde{f}(x,y) = (x, f(x,y))$, respectively ; then theorem 2 gives rise to the desired inequalities. The last statements follow from (18) and the identities :

$$I(\xi \wedge \eta) = I(\xi, f(\xi) \wedge \eta) = I(f(\xi) \wedge \eta) + I(\xi \wedge \eta \;\; f(\xi))$$

$$H(\xi) = H(\xi, f(\xi)) \geqq H(f(\xi))$$

$$H(\xi \mid \eta) = H(\xi \mid \eta, f(\eta)) \leqq H(\xi \mid f(\eta))$$

$$H(\xi \mid \eta) = H(\xi, f(\xi,\eta) \mid \eta) \geqq H(f(\xi,\eta) \mid \eta)$$

respectively.

Theorem 4. (Convexity properties).

Consider the entropy and the mutual information as a function of the distribution of ξ, in the latter case keeping the conditional distributions $\mathcal{P}_{\eta|\xi=x} = \{ p_x(y), y \in Y \}$ fixed :

$$(34) \qquad H(\mathcal{P}) = - \sum_{x \in X} p(x) \log_2 p(x)$$

$$(35) \qquad I(\mathcal{P}) = \sum_{x \in X, y \in Y} p(x)\, p_x(y) \log_2 \frac{p_x(y)}{q_p(y)} ; \quad q_p(y) = \sum_{x \in X} p(x) p_x(y).$$

Then $H(\mathcal{P})$ and $I(\mathcal{P})$ are concave functions of the PD $\mathcal{P} = \{ p(x), x \in X\}$ i.e., if $\mathcal{P}_1 = \{p_1(x) = x \in X\}$, $\mathcal{P}_2 = \{p_2(x), x \in X\}$ and $\mathcal{P} = a\mathcal{P}_1 + (1-a)\mathcal{P}_2 = \{a p_1(x) + (1-a) p_2(x), x \in X\}$ where $0 < a < 1$ is arbitrary, then

$$(36) \qquad H(\mathcal{P}) \geqq a H(\mathcal{P}_1) + (1-a) H(\mathcal{P}_2), \quad I(\mathcal{P}) \geqq a I(\mathcal{P}_1) + (1-a) I(\mathcal{P}_2).$$

Proof. The function $f(t) = -t \log_2 t$ is concave hence so is $H(\mathcal{P})$ as well. Since the PD $\mathcal{Q}_{\mathcal{P}} = \{q_p(y), y \in Y\}$ depends linearly on the PD $\mathcal{P}$, the concavity of $f(t) = -t \log_2 t$ also implies that

$$\sum_{x \in X} p(x)\, p_x(y) \log_2 \frac{p_x(y)}{q_p(y)} =$$

$$= - q_p(y) \log_2 q_p(y) + \sum_{x \in X} p(x) p_x(y) \log_2 p_x(y)$$

is a concave function of $\mathcal{P}$, for every fixed $y \in Y$. Summation for all $y \in Y$ shows that $I(\mathcal{P})$ is concave, too.

Theorem 5. (Useful estimates with the I-divergence).

Let $\mathcal{P}=\{p(x),\ x\in X\}$ and $\mathcal{Q}=\{q(x),\ x\in X\}$ be two PD's on X. Then

$$\sum_{x\in X}|p(x)-q(x)| \leq \sqrt{\frac{2}{\log_2 e}\, I(\mathcal{P}\|\mathcal{Q})} \tag{37}$$

$$\sum_{x\in X} p(x)\left|\log_2\frac{p(x)}{q(x)}\right| \leq I(\mathcal{P}\|\mathcal{Q}) + \min\left(\frac{2\log_2 e}{e},\ \sqrt{2\log_2 e\cdot I(\mathcal{P}\|\mathcal{Q})}\right). \tag{38}$$

Proof. let $A=\{x: p(x)\leq q(x)\}$, $B=\{x: p(x)>q(x)\}$; put $p(A)=p$, $q(A)=q$. Then $p\leq q$, $p(B)=1-p$, $q(B)=1-q$,

$$\sum_{x\in X}|p(x)-q(x)| = 2(q-p), \tag{39}$$

while from (11) and (12) it follows

$$I(\mathcal{P}\|\mathcal{Q}) \geq p\log_2\frac{p}{q} + (1-p)\log_2\frac{1-p}{1-q}. \tag{40}$$

A simple calculation shows that

$$p\log_2\frac{p}{q} + (1-p)\log_2\frac{1-p}{1-q} - 2\log_2 e\cdot(p-q)^2 \geq 0 \qquad (0\leq p\leq q\leq 1) \tag{41}$$

(for $p=q$ the equality holds and the derivative of the left hand side of (41) with respect to p is ≤ 0 if $0<p\leq q<1$).

The relations (39), (40), (41) prove (37).

From (11) and (12) it also follows

$$\sum_{x\in X} p(x)\left|\log_2\frac{p(x)}{q(x)}\right| = I(\mathcal{P}\|\mathcal{Q}) - 2\sum_{x\in A} p(x)\log_2\frac{p(x)}{q(x)} \leq$$

$$\leq I(\mathcal{P}\|\mathcal{Q}) - 2p\log_2\frac{p}{q} = I(\mathcal{P}\|\mathcal{Q}) + 2p\log_2\frac{q}{p}. \tag{42}$$

Here

$$2p\log_2\frac{q}{p} = 2\log_2 e\cdot p\ln\frac{q}{p} \leqq 2\log_2 e\cdot p\ln\frac{1}{p} \leqq \frac{2\log_2 e}{e}$$

(since $f(t) = t\ln\frac{1}{t}$ takes on its maximum for $t=\frac{1}{e}$); furthermore as $\ln\frac{q}{p} = \ln\left(1+\frac{q-p}{p}\right) \leqq \frac{q-p}{p}$, we also have (using (39)) $2p\log_2\frac{q}{p} \leqq 2\log_2 e\cdot(q-p) = \log_2 e\cdot\sum_{x\in X}|p(x)-q(x)|$.

In view of these estimates, (42) and (37) imply (38).

1. Introduction.

In what follows we shall refer to a communication system as schematized by the block-diagram of Fig. 1.1. Such a block-diagram is actually an over-

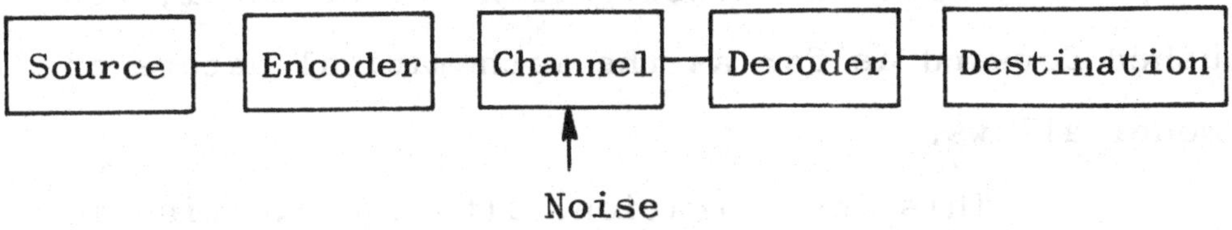

Fig.1.1. The block-diagram for a Communication System

simplification for most of real communication system, but its usefulness has been largely recognized since the pioneering work of Shannon.

This block-diagram has the merit of splitting the system into its main constituent parts, thus permitting one to study these different parts separately, thereafter looking for their possible reciprocal matching. Obviously the matching of the parts is a further problem, whose solution is by no means insured by the knowledge one has of the parts.

Each of the blocks in Fig. 1.1 stands for a mathematical entity, and studying the blocks and their matching amounts to studying such mathematical entities and their relationships. From this point of view what Fig. 1.1 actually represents is a mathematical model for a communication system or for a class of communication systems, and a theoretical study of these systems, i.e. an investigation on the model, can be pushed forward as far as the mathematical nature of the model allows.

This investigation, although exciting in its own merit, could possibly be sterile, were not the choice of the model inspired by practical considerations. And viceversa gaining insight into the theoretical model will be useful for studying and designing real systems. First of all a detailed theoretical study can influence or change our views as what the important parameters of a real communication link are. Moreover the tradeoff between these parameters will be more vividly enlightened, and in turn this will make the construction of detailed mathematical models for real systems easier and closer to reality.

In the diagram of Fig. 1.1 the blocks are not homogeneous. We can think of the Source, Channel and Destination as given, as regards their characteristics, within a narrow band of choices, while there is a much greater flexibility on the characteristics of the Encoder and Decoder. This can depend on the fact that very often the channel is a natural way of communication (the free space, etc.), or that a slight improvement in its characteristics can increase costs dramatically. On the contrary coder and decoder are generally much smaller and less expensive apparatus. Nevertheless, as obvious, there are many exceptions and sometimes it is much more advisable to modify the channel than to improve coding - decoding techniques. E. g. if the channel is Gaussian and white, then its performance can be increased simply by increasing the signal-to-noise ratio, which is a characteristic of the channel and this can improve transmission as much as we like it. Anyway each system has its own characteristics. Stated very roughly, Encoder and Decoder should be designed keeping in mind the characteristics (given) of both the source and the channel on one side and of both the channel and the destination on the other.

What the source really "sees" on its right is not the channel, but the encoder, and what the channel really "sees" on its left is not the source but the encoder. It follows that if the encoder were able to present to the source a handsome channel and to the channel a beautiful source, their coupling would not give rise to any problem. The same of course can be said for channel and destination.

Presented in this way, the task of designing the encoder and the decoder looks rather difficult, and one is led to look for an escape from this uneasy situation. The first thing one thinks of is to split the encoder into two parts : a Source Encoder and a Channel Encoder. Similarly one can think of separating the decoder into a Channel Decoder and a Source Decoder, see Fig.1.2.

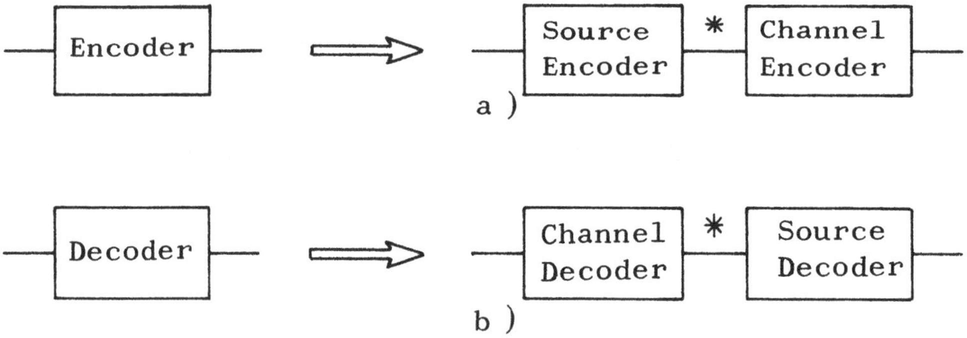

Fig.1.2. Splitting Encoder and Decoder into tow blocks.

In case this separation were possible, it would have the advantage that each of the two blocks, of the encoder say, could be designed independently of the other. The source encoder should take into account the characteristics of the source only, and should permit a "faithful" representation of the source itself in the point marked by a star in Fig. 1.2 a). The channel encoder should take into consideration the characteristics of the channel only and should permit the messages sent through the channel to be "recognizable" at the channel decoder with a small probability of error.

Furthermore, if the splitting indicated in Fig. 1.2 is accomplished, then the same source can feed many channels and the same channel can be fed by many sources, because the matching is now taken at the point marked by a star.

If we want the splitting of encoder and decoder be possible, it is necessary that the source encoder is able to transform the output of a (possibly very broad) class of sources into such a form which in turn the channel encoder should be able to transform into an input acceptable by a (possibly very broad) class of channels.We may think for instance of the information generated by the source as being transformed by the source encoder into a sequence of binary digits. Well, it is by no means obvious that a large class of sources can be represented faithfully enough by such binary sequences, or that binary sequences can always be transformed into something transmittable through channels from a large class. That this is possible, i. e. that such a form for the information generated by the source and to be transmitted through the channel exists, is by no means obvious ; the proof of its existence is rather one of the most important results of Information Theory.

Once the possibility of separating the problems of source-encoding and channel-encoding is ensured, a new problem arises, concerning the optimality of the solution thus provided. It has not been proven that by means of this separation one can achieve an optimal solution for the problem of transmitting with a given source-channel pair ; nevertheless, if the theoretical optimum were not attained, the gain in simplicity of implementation would likely compensate for the loss in optimality.

Moreover, as it looks obvious, this separation of source and channel encoding considerably facilitates the mathematical analysis as well.

In these notes we shall focus on the problem of source encoding, which consists - loosely speaking - in looking for a "faithful" representation of the source, keeping in mind that we want to transmit information reliably and with a minimum number of channel uses.

2. General Description of Information Sources.

As we shall see later, an information source is defined, from a mathematical point of view, as a random process (usually either discrete or continuous).

From a broader, non-formal point of view, an information source could perhaps be defined as a system whatsoever able to interact with other systems ; in this interaction an exchange of energy takes place as well as an exchange of "information". Here by information one could understand what is able to make one system modify its behaviour and/or attitude towards other systems. It seems natural that what is information for one system could well be no information for another system. In many instances the global information (i. e. activity) produced by a system (considered as a source) can be splitted into several parts ; each of which is relevant to a different system (receiver), with an additional possible change in time.

It is not meant here to go into these

general questions thoroughly, partly because the terms involved would need a much more careful definition, which could possibly lead us very far from our aim. It is only important to remark that in many instances of interest the information produced by a source and relevant to a given receiver is or can be expressed in form of a sequence of numbers, or of a numerical function whose values change continuously in time.

In order to evaluate the performance of a communication system (such an evaluation serves as a basis for designing systems or for comparing them), we need some performance criteria, that rest upon some suitable concept of "distortion". Once the distortion is defined, instances of performance criteria are :

1) The maximal distortion (in a given time interval)
2) The average distortion (in a given time interval)
3) The mean value of the square of the distortion (in a given time interval) etc.

Of course the choice of the performance criterion as an index of quality of the system depends upon the specific use we make of the system itself. If an error whatsoever in transmission can have serious consequences even if it happens only once, then 1) is the proper performance criterion. On the other hand, since 1) leads to designing the system for the worst possible case, it is a very expensive criterion, which should therefore be avoided in case an error once in a while does not disturb very much. In this case 2) and 3) are better criteria (i. e. more economical).

Now we want to turn to the description of the source in terms of random processes. Perhaps at this point a remark is in order : while the source, if it has to be meaningful, must generate sensible information in a deterministic way, nevertheless we are going to study its behaviour in a statistical frame. Actually, as it will turn out in the sequel that, far from being contradictory, this approach is very appropriate, as the many interesting results obtained in this way can testify. It will become apparent that only the statistical properties of the source are relevant for designing the coding and decoding devices and for improving the performances of the transmission link. On the contrary, what concerns the real meaning of the particular messages generated does not matter at all. This point of view permits us to design a system able to perform according to the given specifications in front of every particular message whose characteristics fall within the statistical frame of the source. This approach corresponds to statistical performance criteria, of the type 2) and 3) above.

Consider a probability space $(\Omega, \mathcal{F}, p)$ (cfr. preliminaries) and a family W_t of time functions ($t \in \mathcal{T}$, being $\mathcal{T}$ the time index set, which usually is either discrete or continuous) such that to every point $\omega \in \Omega$ there corresponds a function $f_\omega(t) \in W_t$:

$$(2.1) \qquad W_t = \{ f_\omega(t), \ \omega \in \Omega \}, \quad t \in \mathcal{T}.$$

For every fixed $\bar{t} \in \mathcal{T}$ the set $\{ f_\omega(\bar{t}), \ \omega \in \Omega \}$ is a random variable (rv), i. e. the set $B(x, \bar{t}) = \{ \omega : f_\omega(\bar{t}) < x \}$ belongs to $\mathcal{F}$ for every value of x, and there exists the probability of $B(x, \bar{t})$.

Now, suppose the source carries out an experiment, according to the probability distribution p, whose result is a point $\bar{\omega} \in \Omega$. Then the function $f_{\bar{\omega}}(t)$ is generated and considered as the source output.

It is worth remarking at this point that the receiver needs the information generated by the source only with a finite degree of accuracy, agreed upon before starting transmission (the accuracy is measured according to the given distortion measure). This entails that possibly there will be different $f_\omega(t)$ functions which have the same meaning to the receiver, and are thus equivalent. It is therefore convenient to partition the set W_t in (2.1) into "equivalence classes", each of which can be represented by a suitably chosen time function :

$$W_t = \bigcup_{\alpha} V_t^{(\alpha)} , \tag{2.2}$$

$V_t^{(\alpha)}$ an equivalence class

$$f_\alpha(t) \text{ represents the class } V_t^{(\alpha)} . \tag{2.3}$$

The partition (2.2) of W_t into classes is equivalent to the partition of Ω into disjoint sets, which we can label Ω_α :

$$\Omega = \bigcup_{\alpha} \Omega_\alpha . \tag{2.4}$$

Whenever $\omega \in \Omega_\alpha$, the source emits a function $f_\omega(t)$ equivalent to the representative $f_\alpha(t)$ of $V_t^{(\alpha)}$. Consequently, as far as the receiver is concerned, this reduction of the data of interest can be depicted by the introduction of an "equivalent source", whose possible outputs are the representatives $f_\alpha(t)$ (see Fig. 2.1).

In the sequel, when speaking of a source, we shall tacitly refer to this equivalent source, assuming that this reduction has already been accomplished. We shall omit the subscript α for simplicity's sake, and write $f(t)$ instead of $f_\alpha(t)$.

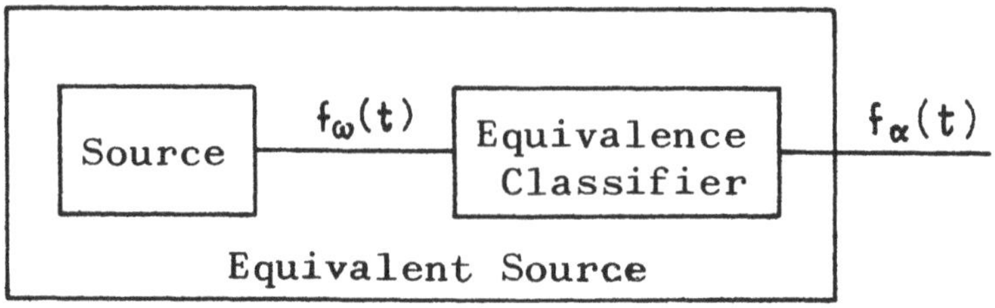

Fig. 2.1. Introduction of the equivalent source.

The purpose of a communication system is to make the function of time $f(t)$, generated by the source, be known at the destination. There are many obstacles to such a job, and the received function $\hat{f}(t)$ will be in general different from $f(t)$:

(2.5) $$\hat{f}(t) \neq f(t).$$

Of course (2.5) is a qualitative statement, that must be given a quantitative character in order to evaluate, and possibly to improve, the system's performance. It is important to remark that an improvement of the performance has a cost, which must be evaluated, and it is the tradeoff between cost and performance that constitutes the important parameter for the system engineer.

3. Discrete Sources.

A Discrete Source (DS) is represented by a discrete random process, i. e. a random process for which the time index set $\mathfrak{J}$ (cfr. (2.1)) is a discrete set of the real axis. In this case each of the functions $f(t)$ produced by the source is a sequence of points belonging to some space(*),or alphabet, $\mathfrak{a} = \{a_1, a_2, \dots\}$. The task of the Source Encoder is to transform (encode) the random sequences that the source can generate into sequences of symbols taken from a given alphabet (code alphabet) $\mathfrak{C} = \{c_1, c_2, \dots\}$ in such a way that 1) the retrieval of the original sequence from the encoded sequence is possible, at least with high probability, and 2) that the number of code symbols required per source symbol is as small as possible.

The model of a DS as depicted here is obviously useful for studying real sources which produce

---

(*)After the reduction to the equivalent source mentioned in 2, the points of this space will be at most countable. A further reduction is always reasonable, and the space can thereafter be considered as finite. To justify this second reduction, many arguments could be alleged (e.g. every concrete system always works with a finite power, etc.).

discrete data, but also for dealing with continuous (waveform) sources whose output has been converted into discrete data(*).

In these notes we shall be concerned only with DS's specified by random sequences. The parallel course held by professor R. G. Gallager will treat the problem of encoding a waveform source into discrete sequences subject to limitations on the distortion.

Fig. 3.1. summarizes the procedure we have been exposing for arriving at a DS starting from any given source. In case the given source is a DS, the block marked "Analog-to-Digital Converter" is absent, and the block marked "Equivalence Classifier" may be absent.

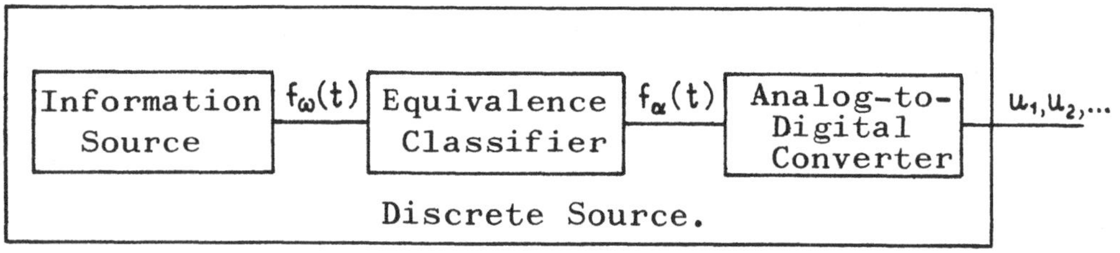

Fig. 3.1. Obtaining a DS from a given source.

(*)The conversion from continuous to discrete form (analog-to-digital conversion) for the source output can be accomplished by means of sampling and quantization (cfr. PCM techniques), which are largely substantiated by the so-called Sampling Theorem.

The output of the Analog-to-Digital converter, and therefore of the whole DS is a sequence $u_1, u_2, \ldots$ of symbols (letters) belonging to the source alphabet $\mathcal{A}$.

From now on it will be immaterial how the DS originated : we shall consider it as a black box emitting sequentially letters from $\mathcal{A}$, and we shall assume that $\mathcal{A}$ contains K symbols :

$$\mathcal{A} = \{a_1, a_2, \ldots, a_K\} \qquad (K \geq 2). \tag{3.1}$$

Every tolerable approximation has been carried out within the black box, and consequently the emitted sequence

$$u_1, u_2, \ldots \qquad (u_i \in \mathcal{A}) \tag{3.2}$$

should be transmitted as carefully as possible.

From a mathematical point of view, the DS of Fig. 3.1 can be described by a probability space $(\Omega, \mathfrak{F}, p)$ on which a sequence of rv's $U_1, U_2, \ldots$ is defined. The DS chooses a point $\bar{\omega} \in \Omega$ and puts out the sequence

$$u_1 = U_1(\bar{\omega}), \; u_2 = U_2(\bar{\omega}), \ldots \qquad (u_i \in \mathcal{A}). \tag{3.3}$$

Therefore, a DS is specified by $\mathcal{A}$ and by the family of probability distributions :

$$P(u_j, u_{j+1}, \ldots, u_{j+n}) \overset{(*)}{\triangleq} P\{\omega : U_j(\omega) = u_j,\ U_{j+1}(\omega) = u_{j+1}, \ldots$$

$$U_{j+n}(\omega) = u_{j+n}\},\ n = 0,1,2,\ldots; \tag{3.4}$$

j an integer.

Remark that in (3.4) we have allowed j to take on also values different from 1, since we are interested in the probability of sequences starting at any instant.

A further specification is needed, concerning the rate at which the symbols are delivered by the source. Actually the rate can be fixed (i. e. the symbols are delivered at regular time intervals) or variable. We shall be concerned only with fixed-rate sources.

The simplest conceivable DS is the so-called "Discrete Memoryless Source" (DMS), for which the rv's $U_1, U_2, \ldots$ are independent and equally distributed. This amounts to say that the source letters are produced according to some fixed probability distribution :

$$P(a_1),\ P(a_2), \ldots P(a_K);\ \left(P(a_i) \geq 0\ ;\ \sum_1^K{}_i\, P(a_i) = 1\right) \tag{3.5}$$

(*)The symbol "$\triangleq$" standing between two terms means that the left-hand term is defined by the right-hand one.

and that the probability of any sequence $u_1, u_2, \dots, u_n$ is given by

$$P(u_1, u_2, \dots, u_n) = P(u_1)P(u_2)\dots P(u_n) = \prod_{i=1}^{n} P(u_i), \qquad (3.6)$$

where

$$P(u_i) \triangleq P\{\omega : U_j(\omega) = u_i\}, \ \forall j, i = 1, 2, \dots, n. \qquad (3.7)$$

Thus a DMS is equivalent to what is called a "finite scheme" :

$$A = \begin{pmatrix} a_1 & a_2 & \dots & a_K \\ p_1 & p_2 & \dots & p_K \end{pmatrix}, \qquad (3.8)$$

being

$$p_i = P(a_i) \qquad (1 \le i \le K). \qquad (3.9)$$

Although DMS's are extreme idealizations for real sources, nevertheless their analysis permits developing many fundamental concepts of source coding, which serve as foundations for more complicated and realistic source models.

4. Variable - Length and Fixed - Length Codes.

According to what we have been exposing so far, it is convenient to design the communication system in such a way that it treats equally every sequence of given length output by the source. In other words, the "meaning" of the sequence will be of no interest for the system that has to transmit it. This implies that the transmission cost is directly proportional to the length of the encoded sequences, i. e. the sequences into which the encoder transforms the source sequences.

Now we introduce the following notations :

(4.1) $$\underline{u}^{(L)} = u_1, u_2, \ldots, u_L \qquad (u_i \in \mathcal{A})$$

will denote a sequence of length L of symbols taken from the source alphabet $\mathcal{A}$.

(4.2) $$\underline{s}^{(N)} = s_1, s_2, \ldots, s_N \qquad (s_i \in \mathcal{C})$$

will denote a sequence of length N of symbols taken from the code alphabet $\mathcal{C}$. The $\underline{s}^{(N)}$ sequences will be called "codewords".

Then what the source encoder actually performs is the following : the original source sequence $u_1, u_2, \ldots$ is broken in a sequence of subsequences of length L of the form (4.1), so that the source is really thought of as emitting a sequence :

(4.3) $$\underline{u}_1^{(L)}, \underline{u}_2^{(L)}, \ldots$$

Then to each of the $\underline{u}_i^{(L)}$ sequences a codeword $\underline{s}_i^{(N_i)} =$

$= \Phi(\underline{u}_i^{(L)})$ is associated whose length N_i depends on the source sequence $\underline{u}_i^{(L)}$: $N_i = N_i(\underline{u}_i^{(L)})$. When the source encoder is fed with the sequence (4.3) it puts out the sequence

$$\Phi(\underline{u}_1^{(L)}), \Phi(\underline{u}_2^{(L)}), \ldots = \underline{s}_1^{(N_1)}, \underline{s}_2^{(N_2)}, \ldots \quad (N_i = N_i(u_i^{(L)})). \quad (4.4)$$

The mapping Φ in (4.4) which defines the codewords to be assigned to the source sequences, is called a "code".

We remark explicitly that our definition of code as given by (4.4) is somewhat restrictive, since the sequences input to the encoder are thought of as having all the same length L. On the other hand we have allowed the codewords to be of different lengths, N_i. Apart from the very special case(*) of a DMS for which $P(a_i) = 1/K$ $(1 \le i \le K)$, the sequences $\underline{u}_i^{(L)}$ output by the source will have different probabilities. (Of course this probability distribution on the L-length sequences is assigned once the source is assigned).

As a consequence, in order to minimize the (average) length of the codewords, and therefore the cost of transmission, it will be convenient to choose a code Φ that associates short codewords to

---

(*) This case will be always excluded from our subsequent considerations as it has no interest.

the most probable source sequences. It is interesting to point out that in many practical instances of coding techniques, such as shorthand and Morse code,this principle ha been applied long before Information Theory was developed.

We do not intend to develop here the theory of variable-length encoding techniques, that will be treated thoroughly by professor G. Katona.

We want only to point out that variable length codes suffer from a serious disadvantage, that we now illustrate. Suppose the source produces its letters at a constant rhythm, say one every σ seconds and the channel transmits the code letters at a constant rhythm as well, say one every τ seconds. Then when the source occasionally emits a very improbable sequence, a very long codeword is sent through the channel, which makes a waiting-line problem arise ; this in turn makes it necessary to insert a buffer between source encoder and channel, for storing the waiting code letters. Of course one could avoid the presence of the buffer keeping-if possible-the rhythm of the channel at a sufficiently high level as to afford stopless transmission. On the other hand, however, such a solution would make the channel not work for most of the time, which is a very poor exploitment of it. Unfortunately, even in case we decide to work with a buffer, if it is of finite size (which is certainly the case in any practical situation) it will overflow with probability 1, thus causing a loss of possibly valuable information.

Two ways are available for coming out of this difficulty : either one endows the buffer with an alarm codeword to be sent whenever the buffer overflows, or one chooses a completely different strategy, using fixed-length encoding techniques.

The rest of these notes will be devoted to fixed-length encoding techniques for fixed-rate sources.

5. An Encoding Scheme for Fixed - Rate Sources.

To sum up things, we wish to encode sequences of length L made up by letters taken from the source alphabet $\mathcal{A}$ (having size K) into codewords of length N made up by letters taken from the code alphabet $\mathcal{C}$ (having size D). We shall consider only DMS's. Codes having the codewords all of the same length are called "block-codes".

Clearly there are K^L different source sequences of length L, while there are D^N different codewords of length N. If we want the code function Φ defined above to have an inverse, then necessarily

(5.1) $$D^N \geq K^L,$$

or equivalently :

(5.2) $$N \geq L \frac{\log_2 K}{\log_2 D},$$

which sets a lower bound on the codeword length. Of course, if condition (5.1) is satisfied, it is possible to encode the L-sequences output by the source in such a way the probability of erroneous decoding, say P_e, is zero. On the other hand, such a presentation of block codes makes them appear as a device for merely translating from alphabet $\mathcal{A}$ to alphabet $\mathcal{C}$. Actually it is possible, using block-codes, to decrease the lower bound on N given by eq. (5.2) while keeping the error probability arbitrarily close to zero. The price one must pay for this decrease in N is that P_e is never exactly zero, although it can be made as small as we wish simply by increasing L. On the other hand, according to what we have said in section 2 about the accuracy requirements of the receiver, there is always a positive probability of error the receiver is ready to tolerate. So, since we can let a sufficiently small P_e enter our encoding-decoding scheme, block codes seem very promising.

Now we look closer at block codes, illustrating a simple encoding scheme which will be very useful also in subsequent developments. If we give up the idea of having an invertible Φ function and provide only W ($W < K^L$) distinct codewords, then there will be codewords coming (a priori) from more than one sequence, which makes decoding ambiguous and introduces a nonzero P_e. It is rather obvious that, once W is fixed, we can make P_e minimum by assigning the W distinct codewords to the W most probable L-length source sequences, while encoding arbitrarily the remaining sequences ; upon reception of any codeword, it should be decoded into the most probable of the source sequences from which it could derive.

Let $\mathcal{B}_L$ be the set of the W most probable L-length source sequences,and $\mathcal{B}_L^c$ its complementary set, which we can call the "ambiguous set". Then obviously we have for the probability of erroneous decoding P_e:

$$P_e = P(\mathcal{B}_L^c), \tag{5.3}$$

being $P(\mathcal{B}_L^c)$ the probability that the source generates a sequence in $\mathcal{B}_L^c$ i. e.

$$P(\mathcal{B}_L^c) \triangleq \sum_{u^{(L)} \in \mathcal{B}_L^c} P(\underline{u}^{(L)}) \tag{5.4}$$

and $P(\underline{u}^{(L)})$ is the probability of the sequence $\underline{u}^{(L)}$ as defined in (3.6). It follows that if the probability of the ambiguous set $\mathcal{B}_L^c$ is sufficiently small, this strategy is acceptable ; moreover the length N of the codewords should satisfy

$$N \geq \frac{\log_2 W}{\log_2 D} \tag{5.5}$$

rather than (5.2), thus showing a reduction. Of course nobody can tell us whether this reduction is significant for some acceptable value of P_e. Before we investigate this crucial point, let us describe more carefully the encoding scheme.

Consider the L - length source sequences arranged in a vertical array according to their decreasing probabilities (see Fig. (5.1.), i. e. suppose $\underline{u}_j^{(L)}$ is above $\underline{u}_I^{(L)}$ when $P(\underline{u}_j^{(L)}) \geq P(\underline{u}_I^{(L)})$.

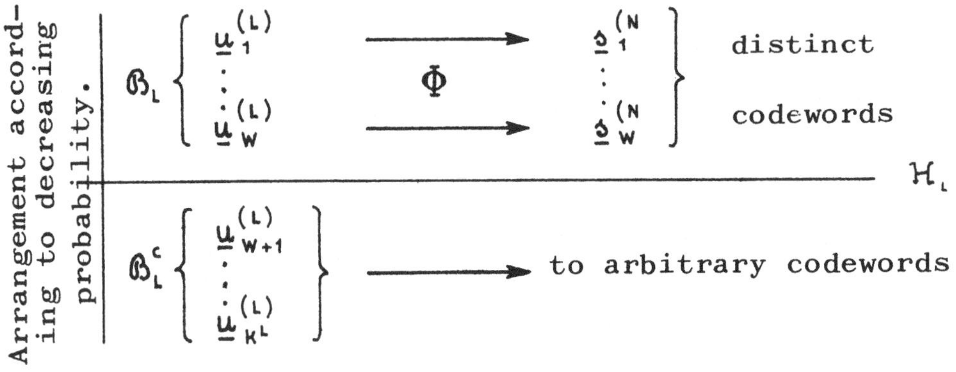

Fig. 5.1. Arrangement of length source sequences according to decreasing probability.

We shall see in the sequel that this

particular ordering of the source sequences will prove to be of outmost importance. Remark that there always exists a suitable number H_L, with $\sup_{\underline{u}_i^{(L)}} P(\underline{u}_i^{(L)}) > H_L > \inf_{\underline{u}_i^{(L)}} P(\underline{u}_i^{(L)})$, such that we can define $\mathcal{B}_L$ as the set of those sequences for which

$$P(\underline{u}^{(L)}) > H_L .$$

Now the problem is the following : is it possible to devise an encoding method which

1) can keep the probability P_e of ambiguous decoding under a prescribed level and

2) can provide a substantial reduction for the block length of the codewords ?

Differently stated (cfr. (5.3) and (5.5)), is it possible to think out an encoding procedure able to provide us with an ambiguous set $\mathcal{B}_L^c$ of "small" probability, and still containing "many" source sequences ?

An answer to this problem is afforded by a theorem of Shannon, which, roughly stated, asserts that to represent the source with an arbitrarily small error probability it is sufficient to provide distinct codewords only for $W \times 2^{LH}$ sequences, being $H = -\sum_1^k p_i \log p_i$ (cfr. (3.9)) the source entropy. Now, since $H < \log K$(*),

(*) This holds if the case $p_i = 1/K$ $(1 \leq i \leq K)$ is excluded.

from (5.5) it follows that N is lower bounded by $LH/\log_2 D$, which is less than $L\log_2 K/\log_2 D$. Actually the ratio $2^{LH}/K^L$ tends to zero when L goes to the infinity, thus bringing about the desired result.

The next section will be devoted to the precise statement and proof of Shannon Theorem.

6. Shannon Theorem on Source Coding.

It is customary in Information Theory to call the quantity

$$I(x) = -\log_2 P(x), \tag{6.1}$$

where $P(x)$ is the probability of the event x, the "self-information" of x. If we look at eq. (3.6), we see that the self-information $I(u_1, u_2, \ldots, u_L)$ of an L-length sequence output by a DMS is the sum of the self-information of its letters :

$$\begin{aligned} I(u_1, \ldots, u_L) &= -\log_2 P(u_1, \ldots, u_L) = -\log \prod_{i=1}^{L} P(u_i) = \\ &= \sum_{1}^{L}{}_i \{-\log_2 P(u_i)\} = \sum_{1}^{L}{}_i I(u_i). \end{aligned} \tag{6.2}$$

Since the single letters in the sequence are independent identically distributed rv's and since, in force of (6.1), self-information is a one-to-one function of probability, eq. (6.2) tells us that the self-information of an L-length sequence output by a DMS is the sum of L independent identically distributed r.v.'s. Then the law of large numbers applies and ensures that, for sufficiently large L the ratio $I(u_1, \ldots, u_L)/L$ tends towards the mean value of the self-information, which coincides with the entropy H; setting $\underline{u}^{(L)} = u_1, \ldots, u_L$ we have thus :

$$\frac{I(\underline{u}^{(L)})}{L} \asymp H \triangleq - \sum_1^K {}_i \, p_i \log p_i \qquad (6.3)$$

(Here we use the symbol $\asymp$ to indicate an approximate equality, that we shall make precise later and that holds asyntotically for large values of L). Taking into consideration the definition (6.1) of self-information, eq. (6.3) can also be written as follows :

$$P(\underline{u}^{(L)}) \asymp 2^{-LH}. \qquad (6.4)$$

Eq. (6.4) expresses an important feature of the set of the K^L sequences of length L output by a DMS : the probability of a "long" sequence is approximately 2^{-LH}, apart from a set of sequences of "negligible" probability. We shall call the sequences of the "probable" set "typical sequences (of length L)". Since the typical sequences (of "large" length L) have an overall probability very close to 1, their number, say M_{typ} is given approximately by

(6.5) $$M_{typ} \asymp 2^{LH}.$$

This makes it possible to encode the typical sequences of length L by means of M_{typ} binary codewords of length $\asymp LH$.

What we have said so far sets in full light the significance of the source entropy H, and relates it, though not precisely for the moment, to the constant H_L introduced in Fig. 5.1 : in a sense, if H_L is smaller than 2^{-LH}, then the typical L-sequences are contained in $\mathcal{B}_L$.

Now we try to be more precise in our statements. First of all let us state precisely the weak law of large numbers, as expressed by eq. (6.3) : for any $\delta > 0$, there exists an $\varepsilon(L, \delta) > 0$ such that

(6.6) $$P\left\{\underline{u}^{(L)} : \left| \frac{I(\underline{u}^{(L)})}{L} - H \right| > \delta \right\} \leq \varepsilon(L, \delta)$$

and

(6.7) $$\lim_{L \to \infty} \varepsilon(L, \delta) = 0.$$

We call "typical sequences" for any fixed L and δ those sequences $\underline{u}^{(L)}$ satisfying the inequality

(6.8) $$\left| \frac{I(\underline{u}^{(L)})}{L} - H \right| \leq \delta.$$

If we label $\mathcal{T}_{L,\delta}$ the set of the typical sequences for any fixed L and δ, then :

$$P\left(\mathcal{T}_{L,\delta}\right) \geq 1 - \varepsilon\left(L,\delta\right). \tag{6.9}$$

The inequality in (6.8) can be put in the following form :

$$L(H-\delta) \leq I\left(\underline{u}^{(L)}\right) \leq L(H+\delta) \tag{6.10}$$

or equivalently

$$2^{-L(H-\delta)} \geq P\left(\underline{u}^{(L)}\right) \geq 2^{-L(H+\delta)} \qquad \left(\underline{u}^{(L)} \in \mathcal{T}_{L,\delta}\right) \tag{6.11}$$

which holds for any typical sequence. Eq. (6.11) is a precise version of eq. (6.4).

We are also in a position to give precise form to eq. (6.5) concerning the number M_{typ} of typical sequences in $\mathcal{T}_{L,\delta}$. First of all, using the right-hand inequality in (6;11) we have :

$$\begin{aligned} 1 \geq P\left(\mathcal{T}_{L,\delta}\right) &\geq M_{typ} \cdot \min_{\underline{u}^{(L)} \in \mathcal{T}_{L,\delta}} P\left(\underline{u}^{(L)}\right) \geq \\ &\geq M_{typ} \cdot 2^{-L(H+\delta)}, \end{aligned} \tag{6.12}$$

whence

$$M_{typ} \leq 2^{L(H+\delta)}. \tag{6.13}$$

On the other hand, from the left-hand inequality in (6.11) :

$$1-\varepsilon(L,\delta) \leq P(\mathcal{T}_{L,\delta}) \leq M_{typ} \cdot \max_{\underline{u}^{(L)} \in \mathcal{T}_{L,\delta}} P(\underline{u}^{(L)}) \leq$$

(6.I4) $$\leq M_{typ} \cdot 2^{-L(H-\delta)},$$

Whence

(6.I5) $$M_{typ} \geq \{1-\varepsilon(L,\delta)\}\, 2^{L(H-\delta)}.$$

Eq. s (6.13) and (6.L5) together yield :

(6.I6) $$\{1-\varepsilon(L,\delta)\}\, 2^{L(H-\delta)} \leq M_{typ} \leq 2^{L(H+\delta)},$$

which is a rigorous statement of eq. (6.5).

If $H_L < 2^{-LH}$ (see Fig. 5.1) then $\mathcal{T}_{L,\delta} \subset \mathcal{B}_L$ and $P(\mathcal{B}_L)$ will tend to 1 as L tends to the infinity. Remark that $\mathcal{B}_L - \mathcal{T}_{L,\delta}$ although possibly containing many sequences (precisely those more probable than $2^{-L(H-\delta)}$ and those whose probability ranges between H_L and $2^{-L(H+\delta)}$) has a probability vanishing as L tends to the infinity.

So what really matters is to provide distinct codewords for the typical sequences. Now if the codeword length N satisfies the inequality

(6.I7) $$N \log_2 D \geq L(H+\delta),$$

then

(6.I8) $$D^N \geq M_{typ}$$

and therefore we can provide one codeword for each typical sequence ; the probability of erroneous decoding is given by

$$P_e = P(\mathcal{T}^c_{L,\delta}) \le \varepsilon(L,\delta). \qquad (6.19)$$

If now L is increased and at the same time N is increased in order to satisfy eq. (6.17), i. e. $N/L \ge (H+\delta)/\log_2 D$, then P_e as given by (6.19) tends to zero for any value of δ . At the same time since $H < \log K$, we are sure that the ratio

$$\frac{M_{typ}}{K^L} \le \frac{2^{L(H+\delta)}}{2^{L\log_2 K}} = 2^{L(H+\delta-\log_2 K)} \qquad (6.20)$$

goes to zero for a suitable choice of δ $(\delta < \log_2 K - H)$.

What we have said in this section gives also a precise meaning the sentence "the source encoder should provide a suitable representation of the source". The adjective "suitable" is now given a rather precise meaning, since the typical sequences are sufficient to represent the source adequately, up to the error probability P_e .

Conversely we want now to prove that the typical sequences are also necessary to describe the source output. In other words, if the ratio N/L is kept at some fixed value below $H/\log_2 D$, the probability of error P_e does not approach zero as L tends to the infinity (this is the so-called weak converse to the source coding theorem). Actually P_e tends to one as L goes to the infinity (this is the so-called strong converse to the source coding theorem).

In fact choose an N satisfying

$$\frac{N}{L} \leq \frac{H-2\delta}{\log_2 D} \tag{6.21}$$

so that the codewords are at most $2^{L(H-2\delta)}$. Since, according to equation (6.11), any typical sequence has a probability not greater than $2^{-L(H-\delta)}$, if we provide distinct codewords for $2^{L(H-2\delta)}$ typical sequences, then

$$2^{-L(H-\delta)} \cdot 2^{L(H-2\delta)} = 2^{-L\delta} \tag{6.22}$$

is the overall probability of the typical sequences for which distinct codewords are provided. One could however feel it appropriate to choose a different strategy : i. e. one could think of letting some typical sequences without codeword, while at the same time providing codewords for some non-typical sequences, namely the most probable among the L-sequences. Since the overall probability of the non-typical sequences does not exceed $\varepsilon(L,\delta)$ (cfr. (6.9)), we can conclude that the total probability of the L-sequences for which we can provide codewords, assuming (6.21) holds, is upperbounded by

$$\varepsilon(L,\delta) + 2^{-L\delta}, \tag{6.23}$$

whence the conclusion that P_e tends to 1 as L goes to the infinity, for any positive δ.

The results we have proven in this section can be summarized as follows :

Theorem 6.1 (Shannon Source Coding Theorem). Let us be given a DMS with entropy H generating L-length sequences of symbols from an alphabet $\mathcal{A}$ of size K. Then whenever

$$N/L \geq (H+\delta)/\log_2 D$$

it is possible to encode these sequences into sequences of length N of symbols from an alphabet of size D in such a way that for any $\varepsilon > 0$ and for sufficiently large L the probability P_e of erroneous decoding satisfies

$$P_e < \varepsilon,$$

whatever the positive constant δ is.

Conversely, no matter what the encoding procedure is, whenever

$$\frac{N}{L} \leq \frac{H-\delta}{\log_2 D}$$

then

$$P_e > 1 - \varepsilon$$

for arbitrary positive ε and sufficiently large L .

We remark that the quantity $\frac{N \log_2 D}{L}$ is apparently of great importance, and it deserves a name of its own. We shall call it the "encoding rate", and label by R :

(6.24) $$R \triangleq \frac{N \log_2 D}{L} = \frac{\log_2 W}{L}$$

where $W = D^N$ is the number of codewords of length N from an alphabet of size D. Therefore the rate R is the ratio of the logarithm of the number of distinct codewords to the length L of the source sequences. If $D=2$, then the equality

(6.25) $$R = N/L$$

shows that R is the number of binary digits required to encode one source digit.

After this definition, theorem (6.1) can also be stated as follows :

For any given rate R greater than H it is possible to encode L-sequences from a DMS into N-sequences with an arbitrarily small error probability P_e , provided only L is sufficiently large. Converse ly, if $R < H$ an arbitrarily small probability of error cannot be achieved, since P_e tends to 1 as L tends to the infinity.

We observe explicitly that the range of interest for R is from 0 to $\log_2 K$:

$$0 < R < \log_2 K$$

since, if $R \geq \log_2 K$, then $D^N \geq K^L$ and a codeword can be provided for each L-sequence, thus making $P_e = 0$ for any value of L, and not only in the limit.

We want to make clear that assigning distinct codewords to the typical sequences is completely equivalent to assigning distinct codewords to a convenient number of the most probable L-sequences. From eq. (6.24) we have for the number W of distinct codewords :

$$W = 2^{LR} \tag{6.26}$$

and once more we must compare R with H. In a precise form we have the following

Theorem 6.2. If one provides distinct codewords for the 2^{LR} most probable L-sequences, then

$$\text{if } R > H \qquad P_e \downarrow 0 \tag{6.27}$$

$$\text{as } L \to \infty$$

$$\text{if } R < H \qquad P_e \uparrow 1 \tag{6.28}$$

Proof. Suppose $R > H$, then

$$2^{-LR} < 2^{-LH}.$$

Setting $m \triangleq \min_{\underline{u}^{(L)} \in \mathcal{B}_L^{(R)}} P(\underline{u}^{(L)})$, where now $\mathcal{B}_L^{(R)}$ is the set of the 2^{LR} most probable L-sequences, we have

$$1 \geq P\left(\mathcal{B}_L^{(R)}\right) \geq 2^{LR} \cdot m,$$

whence

(6.29) $$m \leq 2^{-LR} < 2^{-LH}.$$

In force of (6.29), for sufficiently small δ and large L, $\mathcal{T}_{L,\delta} \subset \mathcal{B}_L^{(R)}$, and therefore $P(\mathcal{B}_L^{(R)}) \uparrow 1$ as $L \to \infty$, thus verifying (6.27).

Conversely, if $R < H$, the set $\mathcal{B}_L^{(R)}$ of the 2^{LR} most probable sequences cannot contain all the typical sequences, which are $\sim 2^{LH} > 2^{LR}$, and consequently for sufficiently large L cannot contain any typical sequence.

To conclude this section, we point out that, although very important, the result expressed by Theorems 6.1 and 6.2 is not very indicative as to the construction and implementation of "good" codes, i. e. of codes capable of the performances foreseen by the theory . Moreover, nothing is said about the speed at which P_e goes to zero as L increases, in case $R > H$, while such a parameter is very important for practical purposes : the complexity of encoding and decoding apparatus grows with L , and an exact estimate of the L needed to achieve a desired probability of error is very useful.

To evaluate such a speed, we now proceed to further investigation.

7. Testing a Simple Alternative.

Consider a probability space $(\Omega , \mathfrak{F})$ and assume that one of two probability distributions μ_1 and μ_2 has been assigned to it, but we do not know which one. To guess at the distribution actually present in $(\Omega, \mathfrak{F})$ we choose a sample (point) x in Ω and this, of course, according to the unknown probability distribution. We divide Ω into two disjoint subsets, E_1 and E_2 $(E_1 \cup E_2 = \Omega ;\ E_1 \cap E_2 = \emptyset)$, and adopt the following decision rule :

1) whenever $x \in E_1$, we say that μ_1 is the true distribution and reject μ_2 ;

2) whenever $x \in E_2$, we say that μ_2 is the true distribution and reject μ_1 .

It is clear that we can be mistaken in two ways : either μ_2 is true and x comes out to belong to E_1 (so that we say wrongly that μ_1 is true) or μ_1 is true and x comes out to belong to E_2 (so that we say wrongly that μ_2 is true). The first mistake is called the "error of the first kind" and has clearly probability $\mu_2(E_1) \hat{=} \gamma$, the second is the "error of the second kind", and has probability $\mu_1(E_2) \hat{=} \beta$.

One could also try to improve the basis on which one takes one's decision, and this can be done e. g. by taking n successive independent samples $x_1, x_2, \ldots, x_n$ from Ω, or equivalently one (n-dimensional) sample from Ω^n. The decision is then made according to whether $(x_1, x_2, \ldots, x_n)$ belongs to some $E_1^{(n)}$ or to its complementary $E_2^{(n)}$ $(E_1^{(n)} \cup E_2^{(n)} = \Omega^n;$ $E_1^{(n)} \cap E_2^{(n)} = \emptyset)$. It should be noted that for each value of n the choice of $E_1^{(n)}$ and $E_2^{(n)}$ ought to lead to a kind of "optimal" decision (e. g. in the sense of minimizing some cost function assigned to the errors of the first and second kind). One rather common problem is the following : for any fixed value $\gamma_0 (0 < \gamma_0 < 1)$ of the error of the first kind, minimize the error of the second kind (by properly choosing $E_1^{(n)}$ and $E_2^{(n)}$), i. e. find out the quantity

$$(7.1) \qquad \beta_n^* \hat{=} \inf_{E_1^{(n)} : \mu_2(E_1^{(n)}) = \gamma_0} \mu_1(E_2^{(n)})$$

or conversely, for any fixed value $\beta_0 (0 < \beta_0 < 1)$ of the error of the second kind, find out the quantity

$$\gamma_n^* \triangleq \inf_{E_2^{(n)} : \mu_1(E_2^{(n)}) = \beta_0} (E_1^{(n)}) . \qquad (7.2)$$

We shall be interested in the behaviour of such quantities as β_n^* and γ_n^* as n goes to the infinity. One helpful tool in this respect is the I-divergence (see Prelimaniries).

Since we are interested in DMS's, we shall restrict our attention to the case of a finite Ω space, say $\Omega = \{a_1, a_2, \dots, a_K\}$; μ_1 and μ_2 are now two K-tuples of numbers :

$$\mu_i = \{\mu_{i1}, \dots, \mu_{iK}\} \qquad (i = 1,2) \qquad (7.3)$$

where

$$\mu_{ij} \geq 0 \quad (1 \leq j \leq K); \quad \sum_1^K {}_j \, \mu_{ij} = 1 \qquad (i = 1,2). \qquad (7.4)$$

The I-divergence of μ_1 and μ_2 is defined as the quantity :

$$I(\mu_1 \| \mu_2) = \sum_1^K {}_j \, \mu_{1j} \log_2 \frac{\mu_{1j}}{\mu_{2j}} \qquad (7.5)$$

and we know that $I(\mu_1 \| \mu_2) \geq 0$, the equality sign holding if and only if μ_1 coincides with μ_2. Of course we have also

$$I(\mu_2 \| \mu_1) = \sum_1^K {}_j \, \mu_{2j} \log_2 \frac{\mu_{2j}}{\mu_{1j}} . \qquad (7.6)$$

The concept of I-divergence can be generalized in a straightforward way to the product distributions of μ_1 and μ_2 over Ω^n, corresponding to the choice of an n-tuple of independent samples from Ω. If we write

$$(7.7) \qquad I(\mu_1 \| \mu_2 ; O_n) \triangleq I(\mu_1^{(n)} \| \mu_2^{(n)})$$

to indicate the I-divergence of the product distribution of μ_1 and the product distribution of μ_2 in Ω^n, it can be easily verified that :

$$(7.8) \qquad I(\mu_1 \| \mu_2 ; O_n) = n\, I(\mu_1 \| \mu_2 ; O_1),$$

where $I(\mu_1 \| \mu_2 ; O_1)$ coincides with $I(\mu_1 \| \mu_2)$.

Actually any n-tuple $\underline{x}^{(n)} = x_1, \ldots, x_n$ of samples from Ω can be considered as a realization of a sequence of independent and identically distributed r.v.'s, each of which can take on K different values, $a_1, a_2, \ldots, a_K$. The μ_i product probability of the n-tuple $\underline{x}^{(n)}$ is

$$(7.9) \qquad \mu_i^{(n)}(\underline{x}^{(n)}) = \mu_i(x_1) \cdots \mu_i(x_n) \qquad (i = 1,2)$$

and the summation over the entire set of the K^n n-length sequences can be performed by summing over a set of n indices $j_1, \ldots, j_n$ each ranging between 1 and K :

$$(7.10) \qquad \sum_{1}^{K}{}_{j_1} \cdots \sum_{1}^{K}{}_{j_n}.$$

So-for the I-divergence of $\mu_1^{(n)}$ and $\mu_2^{(n)}$, we have :

$$I(\mu_1 \| \mu_2 ; O_n) \triangleq I(\mu_1^{(n)} \| \mu_2^{(n)}) = \sum_1^{K^n} {}_j \, \mu_1^{(n)}(\underline{x}_j^{(n)}) \log_2 \frac{\mu_1^{(n)}(\underline{x}_j^{(n)})}{\mu_2^{(n)}(\underline{x}_j^{(n)})} =$$

$$= \sum_1^K {}_{j1} \cdots \sum_1^K {}_{jn} \, \mu_1(x_{j1}) \cdots \mu_1(x_{jn}) \log_2 \frac{\mu_1(x_{j1}) \cdots \mu_1(x_{jn})}{\mu_2(x_{j1}) \cdots \mu_2(x_{jn})} =$$

$$= \sum_1^K {}_{j1} \cdots \sum_1^K {}_{jn} \, \mu_1(x_{j1}) \cdots \mu_1(x_{jn}) \log_2 \frac{\mu_1(x_{j1})}{\mu_2(x_{j1})} +$$

$$+ \cdots \sum_1^K {}_{j1} \cdots \sum_1^K {}_{jn} \, \mu_1(x_{j1}) \cdots \mu_1(x_{jn}) \log_2 \frac{\mu_1(x_{jn})}{\mu_2(x_{jn})} =$$

$$= \sum_1^K {}_{j1} \, \mu_1(x_{j1}) \log_2 \frac{\mu_1(x_{j1})}{\mu_2(x_{j1})} + \cdots + \sum_1^K {}_{jn} \, \mu_1(x_{jn}) \log_2 \frac{\mu_1(x_{jn})}{\mu_2(x_{jn})} =$$

$$= I(\mu_1 \| \mu_2 ; O_1) + \cdots + I(\mu_1 \| \mu_2 ; O_1) = n\, I(\mu_1 \| \mu_2 ; O_1) =$$

$$= n\, I(\mu_1 \| \mu_2) , \tag{7.11}$$

which is eq. (7.8) .

Now we use the concept of I-divergence to study the limiting behaviour of β_n^* and γ_n^* defined by eq.s (7.1) and (7.2). We shall assume $I(\mu_1 \| \mu_2)$ is finite and we shall say that hypothesis h_i holds when μ_i is the true probability distribution $(i = 1, 2)$.

If $x_1, x_2, \ldots x_n$ is our n-tuple sample, then under hypothesis h_1, the quantity

$$\frac{1}{n} \log_2 \frac{\mu_1(x_1) \cdots \mu_1(x_n)}{\mu_2(x_1) \cdots \mu_2(x_n)} = \frac{1}{n} \left\{ \log_2 \frac{\mu_1(x_1)}{\mu_2(x_1)} + \cdots + \log_2 \frac{\mu_1(x_n)}{\mu_2(x_n)} \right\} \tag{7.12}$$

in force of the law of large numbers, tends in probability to the mean value of $\log_2 \frac{\mu_1(x_1)}{\mu_2(x_1)}$ (with respect to the probability distribution μ_1), i. e. to $I(\mu_1 \| \mu_2)$. This means that for arbitrary positive numbers $\varepsilon, \eta, \delta$ and for sufficiently large n we have

$$(7.13) \quad \mathrm{Prob}\left\{ \frac{1}{n} \log_2 \frac{\mu_1(x_1)\ldots\mu_1(x_n)}{\mu_2(x_1)\ldots\mu_2(x_n)} < I(\mu_1 \| \mu_2) - \varepsilon \,\middle|\, h_1 \right\} < \eta$$

$$(7.14) \quad \mathrm{Prob}\left\{ \frac{1}{n} \log_2 \frac{\mu_1(x_1)\ldots\mu_1(x_n)}{\mu_2(x_1)\ldots\mu_2(x_n)} > I(\mu_1 \| \mu_2) + \varepsilon \,\middle|\, h_1 \right\} < \delta$$

Eq. (7.I3) can also be written as follows

$$(7.I5) \quad \mathrm{Prob}\left\{ 2^{n\{I(\mu_1 \| \mu_2) - \varepsilon\}} \leq \frac{\mu_1(x_1)\ldots\mu_1(x_n)}{\mu_2(x_1)\ldots\mu_2(x_n)} \,\middle|\, h_1 \right\} \geq 1-\eta$$

in words : under the hypothesis h_1 the samples of length n split into two disjoint sets, say $E_1^{(n)}$ and $E_2^{(n)}$; the samples in $E_1^{(n)}$ satisfy the inequality

$$(7.I6) \qquad \mu_1(x_1)\ldots\mu_1(x_n) \geq 2^{n(I(\mu_1 \| \mu_2) - \varepsilon)} \mu_2(x_1)\ldots\mu_2(x_n)$$

while the samples in $E_2^{(n)}$ occur with probability less than η, provided n is sufficiently large. Now integrating eq. (7.I6) over $E_1^{(n)}$ we get :

$$1 \geq \text{Prob}\left(E_1^{(n)} \mid h_1\right) \geq 2^{\,n\left(I(\mu_1\|\mu_2)-\varepsilon\right)}\,\text{Prob}\left(E_1^{(n)} \mid h_2\right) \quad / \; (7.17)$$

whence for any fixed value $\beta_0\ (0<\beta_0<1)$ of $\beta=\mu_1(E_2^{(n)})$ i. e. for $\mu_1(E_1^{(n)}) = 1-\beta_0$, taking logarithms in eq. (7.I7) we get :

$$0 \geq \log_2(1-\beta_0) \geq n\left(I(\mu_1\|\mu_2)-\varepsilon\right) + \log_2 \mu_2(E_1^{(n)}), \qquad (7.18)$$

or also

$$\frac{1}{n}\log_2(1-\beta_0) - \frac{1}{n}\log_2 \mu_2(E_1^{(n)}) \geq I(\mu_1\|\mu_2)-\varepsilon \qquad (7.19)$$

and in the limit as n tends to the infinity :

$$\lim_{n\to\infty} \frac{1}{n}\log_2 \frac{1}{\mu_2(E_1^{(n)})} \geq I(\mu_1\|\mu_2) - \varepsilon \;;$$

in particular (see eq. (7.2)) :

$$\lim_{n\to\infty} \frac{1}{n}\log_2 \frac{1}{\gamma_n^*} \geq I(\mu_1\|\mu_2)-\varepsilon \,. \qquad (7.20)$$

Now we state and prove the following theorem :

<u>Theorem 7.1</u> For any fixed value $\beta_0 (0<\beta_0<1)$ of β we have :

$$\lim_{n\to\infty} \frac{1}{n}\log \frac{1}{\gamma_n^*} = I(\mu_1\|\mu_2) \qquad (7.21)$$

which can also be put into the following form :

(7.22) $$\lim_{n \to \infty} (\gamma_n^*)^{\frac{1}{n}} = 2^{-I(\mu_1 \| \mu_2)}$$

Proof. Let $E_3^{(n)}$ be the set of samples of length n satisfying the following inequalities :

(7.23) $$2^{n(I(\mu_1 \| \mu_2)-\varepsilon)} \leq \frac{\mu_1(x_1)\dots\mu_1(x_n)}{\mu_2(x_1)\dots\mu_2(x_n)} \leq 2^{n(I(\mu_1\|\mu_2)+\varepsilon)}$$

on integrating the right-hand inequality in (7.23) over the set $E_3^{(n)}$, we get :

(7.24) $$\text{Prob } (E_3^{(n)}/h_1) \leq 2^{n(I(\mu_1\|\mu_2)+\varepsilon)} \cdot \text{Prob } (E_3^{(n)}/h_2).$$

Now, in force of eq.s (7.13) and (7.14) we have Prob $(E_3^{(n)}/h_1) = \mu_1(E_3^{(n)}) \geq 1-\eta-\delta$, and since, by definition, $E_3^{(n)} \subset E_1^{(n)}$, also Prob $(E_3^{(n)}/h_2) =$

$= \mu_2(E_3^{(n)}) \leq \mu_2(E_1^{(n)}) =$ Prob $(E_1^{(n)}/h_2)$. From (7.24) we get therefore :

$$1-\eta-\delta \leq 2^{n(I(\mu_1\|\mu_2)+\varepsilon)}\mu_2(E_1^{(n)}),$$

whence for any fixed $\eta = \beta_0$ also

(7.25) $$1-\beta_0-\delta \leq 2^{n(I(\mu_1\|\mu_2)+\varepsilon)}\gamma_n^*.$$

Taking logarithms on both sides of eq. (7.25) and passing to the limit as n goes to the infinity :

$$\lim_{n \to \infty} \frac{1}{n} \log_2 \frac{1}{\gamma_n^*} \leq I(\mu_1 \| \mu_2) + \varepsilon . \qquad (7.26)$$

Now combining eq.s (7.26) and (7.20) we get the desired result expressed by eq. (7.21) in force of the arbitrariness of ε .

In the same way one can prove the following

Theorem 7.2. For any fixed value $\gamma_0 (0 < \gamma_0 < 1)$ of γ, the limiting behaviour of β_n^* is described by the following expression :

$$\lim_{n \to \infty} \frac{1}{n} \log_2 \frac{1}{\beta_n^*} = I(\mu_2 \ \mu_1), \qquad (7.27)$$

or

$$\lim_{n \to \infty} (\beta_n^*)^{\frac{1}{n}} = 2^{-I(\mu_2 \| \mu_1)}. \qquad (7.28)$$

8. The Neyman - Pearson Lemma.

In this section we shall state and prove a particular case of the well-known Neyman-Pearson lemma, which gives a useful hint concerning how to choose the sets $E_1^{(n)}$ and $E_2^{(n)}$ in order to achieve the infima β_n^* and γ_n^* of the errors of the second or of the first kind (see definitions (7.1) and (7.2)).

Theorem 8.1. If $E_*^{(n)} \subset \Omega$ is a set such that $\mu_1(E_*^{(n)}) = 1 - \beta_0 (0<\beta_0<1)$ and if $\underline{v} \notin E_*^{(n)}$ implies :

$$(8.1) \qquad \frac{\mu_2(\underline{v})}{\mu_1(\underline{v})} \geq \sup_{\underline{u} \in E_*^{(n)}} \frac{\mu_2(\underline{u})}{\mu_1(\underline{u})}$$

being μ_1 and μ_2 two probability distributions on the finite set Ω, then

$$(8.2) \qquad \gamma_n^* = \mu_2(E_*^{(n)}).$$

Similarly, if $E_{**}^{(n)} \subset \Omega^n$ is a set such that $\mu_2(E_{**}^{(n)}) = 1 - \gamma_0$ and if $\underline{v} \notin E_{**}^{(n)}$ implies

$$(8.3) \qquad \frac{\mu_1(\underline{v})}{\mu_2(\underline{v})} \geq \sup_{\underline{u} \in E_{**}^{(n)}} \frac{\mu_1(\underline{u})}{\mu_2(\underline{u})}$$

then

$$(8.4) \qquad \beta_n^* = \mu_1(E_{**}^{(n)}).$$

Proof. Consider any set $E^{(n)} \subset \Omega^n$ such that $\mu_1(E^{(n)}) \geq 1-\beta_0$. What we wish to prove is that $\mu_2(E^{(n)}) \geq \mu_2(E_*^{(n)})$. Now, since

$$(8.5) \qquad \begin{aligned} E^{(n)} &= (E^{(n)} \cap E_*^{(n)}) \cup (E^{(n)} - E_*^{(n)}) \\ E_*^{(n)} &= (E_*^{(n)} \cap E^{(n)}) \cup (E_*^{(n)} - E^{(n)}) \end{aligned}$$

the two equalities

$$(8.6) \qquad \mu_2(E^{(n)}) \geq \mu_2(E_*^{(n)}) \text{ and } \mu_2(E^{(n)} - E_*^{(n)}) \geq \mu_2(E_*^{(n)} - E^{(n)})$$

are equivalent. So it is sufficient that we prove the right-hand inequality in (8.6). The following chain of equalities and inequalities holds :

$$\mu_2(E^{(n)}-E_*^{(n)}) = \sum_{\underline{v}\in E^{(n)}-E_*^{(n)}} \mu_2(\underline{v}) = \sum_{\underline{v}\in E^{(n)}-E_*^{(n)}} \frac{\mu_2(\underline{v})}{\mu_1(\underline{v})}\,\mu_1(\underline{v}) \geq$$

$$\geq \sum_{\underline{v}\in E^{(n)}-E_*^{(n)}} \left(\sup_{\underline{u}\in E_*^{(n)}} \frac{\mu_2(\underline{u})}{\mu_1(\underline{u})}\right)\mu_1(\underline{v}) = \left(\sup_{\underline{u}\in E_*^{(n)}} \frac{\mu_2(\underline{u})}{\mu_1(\underline{u})}\right)\cdot\mu_1(E^{(n)}-E_*^{(n)}) \geq$$

$$\overset{\S}{\geq} \left(\sup_{\underline{u}\in E_*^{(n)}} \frac{\mu_2(\underline{u})}{\mu_1(\underline{u})}\right)\mu_1(E_*^{(n)}-E^{(n)}) = \sum_{\underline{v}\in E_*^{(n)}-E^{(n)}} \left(\sup_{\underline{u}\in E_*^{(n)}} \frac{\mu_2(\underline{u})}{\mu_1(\underline{u})}\right)\cdot\mu_1(\underline{v}) \geq$$

$$\geq \sum_{\underline{v}\in E_*^{(n)}-E^{(n)}} \frac{\mu_2(\underline{v})}{\mu_1(\underline{v})}\cdot\mu_1(\underline{v}) = \mu_2\left(E_*^{(n)}-E^{(n)}\right),$$

being the inequality marked by (§) a consequence of eq.s (8.7) and of the assumption $\mu_1(E^{(n)}) \geq 1-\beta_0 = \mu_1(E_*^{(n)})$. So the right-hand inequality in (8.6) is proven, which proves the left-hand inequality, and the first part of the theorem, i.e. eq. (8.2). The second part is proven in the same way.

Fig. 8.1 gives an intuitive idea of the theorem we have just proved.

set of all
n - length se-
quences of ele-
ments from Ω

arrangement ac-
cording to de-
creasing $\frac{\mu_2}{\mu_1}$ ratio

$\mu_1\left(E_*^{(n)c}\right) = \beta_0$

$\mu_2\left(E_*^{(n)}\right) = \gamma_n^*$

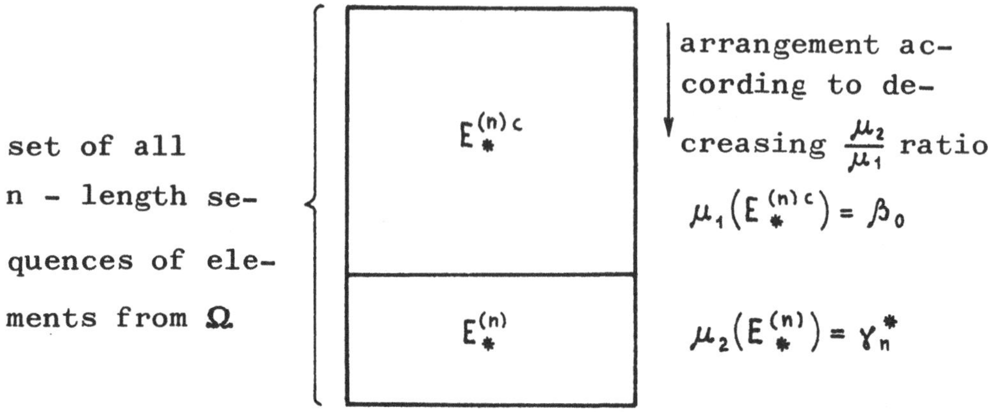

Fig. 8.1.   Illustrating Neyman-Pearson lemma.

9. Proof of Shannon Theorem on Source Coding Based on Neyman - Pearson Lemma.

When we are confronted with the source coding problem, it may happen we know the maximum pro-bability of erroneous decoding, say P_e, the receiver is prepared to tolerate. Of course, to keep transmis-sion costs at a minimum, we want to use the shortest codeword length compatible with that P_e, i. e. we want to use the smallest number of codewords. We have al-ready seen that this minimum number is given asymptot-ically by 2^{LH}, being H the source entropy and L the length of the source sequences.

We want now to prove the same result making use of the Neyman-Pearson lemma. Consider theorem 8.1 and put $n = L$, $\beta_0 = P_e$, $\mu_1 = \mathcal{P} \triangleq \{p_1, \dots, p_K\}$ (i. e. μ_1 coincides with the probability distribution assigned over the alphabet $\mathcal{A}$ of the DMS), $\mu_2 = \mathcal{U} \triangleq \triangleq \{1/K, \dots, 1/K\}$ (i. e. μ_2 is the uniform probability distribution over $\mathcal{A}$). Under these assumptions the set $E_*^{(n)}$ of theorem 8.1 becomes the set $E_*^{(L)}$ for which 1) $P(E_*^{(L)}) = 1 - P_e$ and 2) $\underline{v} \notin E_*^{(L)}$ implies $\left(\frac{1}{K}\right)^L \frac{1}{P(\underline{v})} \geq \sup_{\underline{u} \in E_*^{(L)}} \left(\frac{1}{K}\right)^L \frac{1}{P(\underline{u})}$ i. e. $P(\underline{v}) \leq \inf_{\underline{u} \in E_*^{(L)}} P(\underline{u})$. In other words $E_*^{(L)}$ is the set of the most P-probable sequences of length L whose overall probability is $1 - P_e$.

Since when μ_2 coincides with $\mathcal{U}$ the measure $\mu_2(E^{(L)})$ of a set $E^{(L)}$ is proportional to the number of sequences contained in $E^{(L)}$, say $N(E^{(L)})$, in case we find the minimum of $\mathcal{U}(E^{(L)})$, we have also found the minimum for that number. Actually we have

$$\mu_2(E^{(L)}) = \mathcal{U}(E^{(L)}) = \sum_{\underline{u} \in E^{(L)}} \mathcal{U}(\underline{u}) = \sum_{\underline{u} \in E^{(L)}} \left(\frac{1}{K}\right)^L = N(E^{(L)}) \cdot \left(\frac{1}{K}\right)^L.$$

Now eq. (8.2) gives us

$$\gamma_n^* = \left(\frac{1}{K}\right)^L \cdot N\left(E_*^{(L)}\right) \tag{9.1}$$

and since $\gamma_n^* \cong 2^{LI(\mathcal{P} \| \mathcal{U})}$, we have from (9.1)

$$N\left(E_*^{(L)}\right) \cong K^L \cdot 2^{-LI(\mathcal{P} \| \mathcal{U})}. \tag{9.2}$$

On the other hand, from the very definition of I-divergence we get :

$$I(\mathcal{P}\|\mathcal{U}) = \sum_{1}^{K}{}_i \, p_i \log_2 \frac{1}{K p_i} = H + \log_2 K$$

and substituting into (9.2)

$$N(E_*^{(L)}) \cong K^L \cdot 2^{LH} \cdot 2^{-L\log_2 K}$$

i. e.

$$N(E_*^{(L)}) \cong 2^{LH},$$

which is Shannon theorem.

10. The Auxiliary Distribution.

Now we wish to study the asymptotic behaviour of P_e as L goes to the infinity. To this end, we shall introduce, along with the original probability distribution $\mathcal{P}$ of the DMS a family of auxiliary probability distributions on $\mathcal{A}$:

(10.1) $$\mathcal{Q}_\alpha = \{q_{\alpha 1}, \ldots, q_{\alpha K}\}$$

where :

(10.2) $$q_{\alpha i} = \frac{p_i^\alpha}{\sum_{1}^{K}{}_j \, p_j^\alpha} \qquad (1 \le i \le K)$$

being α a positive real number to be specified later.

Consider now the function of

$$H(\alpha) = -\sum_{1}^{K}{}_{i}\, q_{\alpha i} \log_2 q_{\alpha i}, \tag{10.3}$$

which is nothing else than the entropy of the auxiliary distribution $\mathcal{Q}_\alpha$. $H(\alpha)$ is apparently a continuous function of α for $\alpha > 0$; moreover, it is a decreasing function of α. Actually its first derivative is

$$H'(\alpha) = \alpha \left[\left(\frac{\sum_{1}^{K}{}_{j}\, p_j^\alpha \log_2 p_j}{\sum_{1}^{K}{}_{j}\, p_j^\alpha} \right)^2 - \frac{\sum_{1}^{K}{}_{j}\, p_j^\alpha (\log_2 p_j)^2}{\sum_{1}^{K}{}_{j}\, p_j^\alpha} \right] \tag{10.4}$$

and for $\alpha > 0$ $H'(\alpha)$ is a negative function, as one can see by multiplying the terms in square brackets by $(\sum_{1}^{K}{}_{j}\, p_j^\alpha)^2$, which yields

$$\left(\sum_{1}^{K}{}_{j}\, p_j^\alpha \log_2 p_j \right)^2 - \sum_{1}^{K}{}_{j}\, p_j^\alpha \cdot \sum_{1}^{K}{}_{j}\, p_j^\alpha$$

and by applying Cauchy inequality

$$\left(\sum_{1}^{K}{}_{j}\, a_j b_j \right)^2 \leq \sum_{1}^{K}{}_{j}\, a_j \sum_{1}^{K}{}_{j}\, b_j \tag{10.5}$$

valid for any two sets of K non-negative numbers $\{a_j\}$ and $\{b_j\}$, with $a_j = \sqrt{p_j^\alpha}$ and $b_j = \sqrt{p_j^\alpha} \log_2 p_j$.

Remark that in (10.5) the equality sign holds if and only if $a_j = t\, b_j$ $(1 \leq j \leq K)$ for some positive constant t and consequently $H'(\alpha)$ is zero if and only if $\sqrt{p_j^\alpha} = \sqrt{p_j^\alpha} \log_2 p_j$ $(1 \leq j \leq K)$, i.e. $p_j = \frac{1}{K}$ $(1 \leq j \leq K)$

but we have excluded this case, and therefore $H'(\alpha)$ is strictly negative and $H(\alpha)$ is a strictly decreasing function of α

It is easily seen that

(10.6) $$\lim_{\alpha \to 0} H(\alpha) = \log_2 K$$

(10.7) $$H(1) = H$$

(10.8) $$\lim_{\alpha \to +\infty} H(\alpha) = \log_2 r,$$

being r in eq. (10.8) the number of indices j's for which p_j has its greatest value :

$$p_1 = p_2 = \ldots = p_r > p_{r+1} \geq \ldots \geq p_K .$$

It can be also proved that $H(\alpha)$ is a convex $\cup$ function, so that its behaviour is roughly depicted in Fig. 10.1.

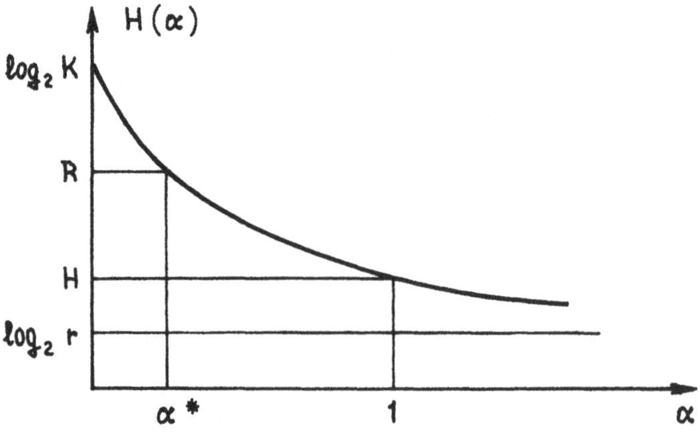

Fig. 10.1. Graph of the function $H(\alpha)$.

It immediately follows that whenever a number R satisfies the limitations :

$$\log_2 r < R < \log_2 K \tag{10.9}$$

there will be one and only one number α, say α^*, such that the equation

$$H(\alpha) = R \tag{10.10}$$

is satisfied. Moreover, in case R lies between H and $\log K$, then α^* lies between 0 and 1. If $\log r < R < H$, then α^* is greater than 1.

We close this section by remarking the following properties of the $\mathcal{Q}_\alpha$ auxiliary distributions:

1) whenever $p(\underline{u}^{(L)}) \geq p(\underline{v}^{(L)})$, then $q_\alpha(\underline{u}^{(L)}) \geq q_\alpha(\underline{v}^{(L)})$ for any positive α ;

2) whenever $p(\underline{u}^{(L)}) \geq p(\underline{v}^{(L)})$, then $\frac{p(\underline{u}^{(L)})}{q_\alpha(\underline{u}^{(L)})} \geq \frac{p(\underline{v}^{(L)})}{q_\alpha(\underline{v}^{(L)})}$ for $0 < \alpha < 1$, while $\frac{p(\underline{u}^{(L)})}{q_\alpha(\underline{u}^{(L)})} \leq \frac{p(\underline{v}^{(L)})}{q_\alpha(\underline{v}^{(L)})}$ for $\alpha > 1$.

To check these properties, we simply remark that $p(\underline{u}^{(L)}) = \prod_1^L {}_i\, p(u_i)$ and $q_\alpha(\underline{u}^{(L)}) = \prod_1^L {}_i \frac{p^\alpha(u_i)}{\sum_1^K {}_j\, p^\alpha(a_j)}$ and that if $t > 0$, then x^t is an increasing function of x.

Properties 1)and 2) above entail that if we order the L-length sequences of letters from the DMS alphabet $\mathcal{A}$ according to decreasing $\mathcal{P}$-probability, then they are ordered according to decreasing $\mathcal{Q}_\alpha$-probability too for each positive α, and according to decreasing $\mathcal{P}/\mathcal{Q}_\alpha$ ratio for any α between zero and one.

11. The Error Exponent.

Suppose a DMS is assigned having an entropy H and a finite alphabet $\mathcal{A} = \{a_1, a_2, \ldots, a_K\}$ with a probability distribution $\mathcal{P} = \{p_1, \ldots, p_K\}$ on it. If an encoding rate R is given, with $R > H$, let α^* be the unique solution to equation (10.10); then α^* lies between 0 and 1, and for a sufficiently small but otherwise arbitrary ε the following inequalities hold :

(11.1) $$H(\alpha^* - \varepsilon) > R > H(\alpha^* + \varepsilon),$$

in force of the decreasing character of the function $H(\alpha)$. Consider the auxiliary distributions $\mathcal{Q}_{\alpha^*-\varepsilon}$ and $\mathcal{Q}_{\alpha^*+\varepsilon}$ defined as in eq. (10.1) and let $\mathcal{B}_L^{(R)}$ be the set of the 2^{LR} most $\mathcal{P}$-probable L-length source sequences. In force of property 1) of section 10, $\mathcal{B}_L^{(R)}$ is also the set of the 2^{LR} most $\mathcal{Q}_\alpha$-probable sequences.

As a direct consequence of Shannon theorem, inequalities (11.1) imply

(11.2) $$\mathcal{Q}_{\alpha^*-\varepsilon}\left(\mathcal{B}_L^{(R)c}\right) \uparrow 1$$

as $L \rightarrow \infty$

(11.3) $$\mathcal{Q}_{\alpha^*+\varepsilon}\left(\mathcal{B}_L^{(R)c}\right) \downarrow 0$$

being $\mathcal{B}_L^{(R)c}$ the complement of $\mathcal{B}_L^{(R)}$.

From (11.2) it follows that for sufficiently large L the set $\mathcal{B}_L^{(R)c}$ satisfies the inequality

$$\mathcal{Q}_{\alpha^*-\varepsilon}\left(\mathcal{B}_L^{(R)c}\right) \geq 1-\gamma$$

for any fixed γ between zero and one, and therefore

$$\inf_{E^{(L)} \subset \mathcal{Q}^L : \mathcal{Q}_{\alpha^*-\varepsilon}(E^{(L)}) \geq 1-\gamma} P\left(E^{(L)}\right) \leq P\left(\mathcal{B}_L^{(R)c}\right) \qquad (11.4)$$

On the other hand, in force of (11.3), if L is large enough $\mathcal{B}_L^{(R)}$ does not satisfy the inequality

$$\mathcal{Q}_{\alpha^*+\varepsilon}\left(\mathcal{B}_L^{(R)c}\right) \geq 1-\gamma$$

An application of the Neyman-Pearson lemma with $\mu_2 = \mathcal{P}$, $\mu_1 = \mathcal{Q}_{\alpha^*+\varepsilon}$ and $E_*^{(L)} = \mathcal{B}_L^{(R)c}$ shows that

$$\inf_{E^{(L)} \subset \mathcal{Q}^L : \mathcal{Q}_{\alpha^*+\varepsilon}(E^{(L)}) \geq 1-\gamma} P\left(E^{(L)}\right) \geq P\left(\mathcal{B}_L^{(R)c}\right). \qquad (11.5)$$

Applying theorem 7.1 to the infima appearing in eq.s (11.4) and (11.5) we can deduce

$$\begin{aligned} -I\left(\mathcal{Q}_{\alpha^*+\varepsilon} \,\|\, \mathcal{P}\right) &\leq \underline{\lim}_{L\to\infty} \frac{1}{L} \log_2 P\left(\mathcal{B}_L^{(R)}\right) \leq \\ &\leq \overline{\lim}_{L\to\infty} \frac{1}{L} \log_2 P\left(\mathcal{B}_L^{(R)}\right) \leq -I\left(\mathcal{Q}_{\alpha^*+\varepsilon} \,\|\, \mathcal{P}\right) \end{aligned} \qquad (11.6)$$

where $P\left(\mathcal{B}_L^{(R)c}\right)$ is actually the probability of incorrect decoding P_e.

Remark that ε in (11.6) is arbitrarily

small and that $I(\mathcal{Q}_\alpha \| \mathcal{P})$ is a continuous function of α; thus we have proven the following

Theorem 11.1. If the encoding rate R is greater than the entropy H of a DMS operating according to the probability distribution $\mathcal{P} = \{p_1, \dots, p_K\}$, then the limiting behaviour of the probability of incorrect decoding P_e is as follows :

$$(11.7) \qquad \lim_{L \to \infty} \frac{1}{L} \log_2 P_e = - I(\mathcal{Q}_{\alpha^*} \| \mathcal{P})$$

or also

$$(11.8) \qquad P_e \cong 2^{-L I(\mathcal{Q}_{\alpha^*} \| \mathcal{P})}$$

where $I(\cdot \| \cdot)$ is an I-divergence and $\mathcal{Q}_\alpha$ and α^* are defined in (10.2) and (10.10) respectively.

It is important to observe that if $R<H$, then eq. (10.10) still has a unique solution α^*, but now α^* is greater than 1, and therefore, in force of property (10.12), ordering the L-sequences according to decreasing $\mathcal{P}$-probability corresponds to ordering them according to increasing $\mathcal{P}/\mathcal{Q}_{\alpha^*}$ ratio. This entails that in the application of the Neyman-Pearson lemma referred to in eq. (11.5), $\mathcal{B}_L^{(R)c}$ cannot play the role of $E_*^{(L)}$; it is rather $\mathcal{B}_L^{(R)}$ which plays that role. Since in force of eq. (11.1), still valid, we have :

$$\left.\begin{array}{l}\mathcal{Q}_{\alpha^*-\varepsilon}(\mathcal{B}_L^{(R)}) \uparrow 1 \\ \mathcal{Q}_{\alpha^*-\varepsilon}(\mathcal{B}_L^{(R)}) \downarrow 0\end{array}\right. \quad \text{as } L \text{ tends to } \infty$$

we can say that the set $\mathcal{B}_L^{(R)}$ satisfies the inequality $\mathcal{Q}_{\alpha^*-\varepsilon}(\mathcal{B}_L^{(R)}) \geq 1-\gamma$ for any γ between zero and one and for sufficiently large L, so that

$$\inf_{E^{(L)} \subset \mathcal{A}^L : \mathcal{Q}_{\alpha^*-\varepsilon} \geq 1-\gamma} P(E^{(L)}) \geq P(\mathcal{B}_L^{(R)})$$

and on the other side, since the inequality $\mathcal{Q}_{\alpha^*+\varepsilon}(\mathcal{B}_L^{(R)}) \geq$ $\geq 1-\gamma$ cannot be satisfied for arbitrarily large L, applying the Neyman-Pearson lemma yields :

$$\inf_{E^{(L)} \subset \mathcal{A}^L : \mathcal{Q}_{\alpha^*+\varepsilon}(\mathcal{B}_L^{(R)}) \geq 1-\gamma} P(E^{(L)}) \geq P(\mathcal{B}_L^{(R)})$$

Now we remark that $P(\mathcal{B}_L^{(R)}) = 1 - P_e$, and much in the same way as we obtained theorem 11.1, we obtain the following

<u>Theorem 11.2</u>. If the encoding rate R is less than the entropy H of a DMS, then with the same notations as in theorem 11.1 the limiting behaviour of the probability of erroneous decoding P_e is as follows :

$$\lim_{L \to \infty} \frac{1}{L} \log_2 (1 - P_e) = - I(\mathcal{Q}_{\alpha^*} \| \mathcal{P})$$

or also

$$P_e \cong 1 - 2^{-L I(\mathcal{Q}_{\alpha^*} \| \mathcal{P})}. \qquad (11.10)$$

12. Sharpening of the Preceding Results.

Theorem 7.1 of section 7 can be generalized and considerably precised. Since the case of interest for us will be that of a DMS with finite alphabet, we shall restrict the statement of the generalized theorem to the case of a finite set Ω, say $\Omega = \{x_1, \ldots, x_K\}$.

Let $\mu_1 = \{p'_1, \ldots, p'_K\}$ be a probability distribution on Ω and let $\mu_2 = \{a_1, \ldots, a_K\}$ be a set of positive numbers corresponding to the elements of Ω (μ_1 is obviously absolutely continuous with respect to μ_2).

For any L-length sequence $\underline{u}^{(L)} = u_1, \ldots, u_L$ ($u_i \in \Omega$) we put :

$$\mu_1(\underline{u}^{(L)}) = \prod_{i=1}^{L} p'(u_i) ; \quad \mu_2(u^{(L)}) = \prod_{i=1}^{L} a(u_i) \tag{12.1}$$

and for any subset $E^{(L)}$ of Ω^L we put

$$\mu_i(E^{(L)}) = \sum_{\underline{u}^{(L)} \in E^{(L)}} \mu_i(\underline{u}^{(L)}) \qquad (i = 1, 2) . \tag{12.2}$$

Then the following theorem holds :

<u>Theorem 12.1</u>. The exact asymptotic expression for γ_L^* (defined by eq. (7.2) in the particular case when μ_2 is a probability distribution), whenever a fixed value

$\beta_0 (0 < \beta_0 < 1)$ is assigned to β is given by :

$$\log \gamma_L^* = LM + \sqrt{L}\,\lambda S - \frac{1}{2}\log L + \frac{T^3}{6S^2}(\lambda^2 - 1) - \frac{1}{2}\lambda^2 - $$
$$- \log\left(\sqrt{2\pi}\; S\right) + o(1), \qquad (12.3)$$

being M the expectation, S^2 the variance and T^3 the third central moment of the r.v. $h(x_i) \hat{=} - \log \frac{p_i'}{a_i}$ with respect to the probability distribution μ_1, and λ is defined by

$$\int_{-\infty}^{\lambda} \frac{1}{\sqrt{2\pi}}\; e^{-t^2/2}\, dt = 1 - \beta_0. \qquad (12.4)$$

We remark that this theorem, whose proof will be omitted, generalizes theorem 7.1 in that μ_2 is not assumed to be a probability distribution, but simply a (finite) measure on Ω. Of course the validity of theorem 12.1. requires the existence of the moments up to the third order of the r.v. $h(x_i)$, but this requirement is certainly met in the case of a finite set Ω .

Now we wish to use theorem 12.1 to sharpen theorem 11.1 (and at the same time also theorem 11.2) concerning the limiting behaviour of the probability of erroneous decoding P_e. The improved result is expressed by the following

<u>Theorem 12.2</u>. In case the encoding rate R is greater than the entropy H of a DMS, then $\log P_e$ has the fol-

lowing expression :

$$\log_2 P_e = - L\, I\,(\mathcal{Q}_{\alpha^*} \| \mathcal{P}) - \frac{1}{2\alpha^*}\log_2 L + \frac{(1-\alpha^*)T_1^3}{6\alpha^* S_1^2} - \frac{T_2^3}{6 S_2^2} -$$

$$(12.5) \qquad - \frac{1-\alpha^*}{2\alpha^*}\log_2 S_1^2 - \frac{1}{2}\log_2 S_2^2 - \frac{1}{\alpha^*}\log_2 \frac{2\pi}{\log_2 e} + o(1)$$

where $\mathcal{Q}_\alpha$ is the auxiliary distribution defined by (10.2), α^* is the unique solution of eq. (10.10), S_1^2 and S_2^2 are the variances and T_1^3 and T_2^3 are the third central moments of the r.v.s $h_1(x_i) = -\log q_{\alpha^* i}$ and $h_2(x_i) = -\log \frac{q_{\alpha^* i}}{p_i}$ respectively, with respect to the auxiliary distribution $\mathcal{Q}_{\alpha^*}$.

<u>Proof</u>. We refer once more to the set of all L-length source sequences $\underline{u}^{(L)}$ arranged, as in Fig. 5.1, according to decreasing $\mathcal{P}$-probability. Let $\mathcal{B}_L^{(R)}$ denote as usual the set of all the 2^{LR} most $\mathcal{P}$-probable sequences, and choose an α_L such that

$$(12.6) \qquad \mathcal{Q}_{\alpha_L}(\mathcal{B}_L^{(R)}) = \frac{1}{2}.$$

Such a choice for α_L is always possible, provided L is sufficiently large, as it is seen from eq.s (11.2) and (11.3) and from the continuity of $\mathcal{Q}_\alpha(\mathcal{B}_L^{(R)})$ as a function of α .

Now we apply theorem 12.1 twice, setting $\beta_0 = \frac{1}{2}$, which, in view of eq. (12.4), amounts to set

$\lambda = 0$. In the first application we set :

$$p_i' = q_{\alpha_L i} \quad (1 \leq i \leq K), \text{ i.e. } \mu_1 = \mathcal{Q}_{\alpha_L}$$
$$a_i = 1 \quad (1 \leq i \leq K), \text{ i.e. } \mu_2 = \{1, \ldots, 1\}, \tag{12.7}$$

so that M coincides with the entropy $H(\alpha_L)$ of $\mathcal{Q}_{\alpha_L}$. In force of the Neyman-Pearson lemma, the infimum γ_L^* (which is now the minimum number of sequences, since $a_i = 1$ for all i's) is achieved by the set $\mathcal{B}_L^{(R)}$ and therefore $\gamma_L^* = 2^{RL}$, i.e. $\log_2 \gamma_L^* = LR$, so that eq. (12.3) gives

$$LR = L H(\alpha_L) - \frac{1}{2} \log_2 L - \frac{T_1^3}{6 S_1^2} - $$
$$- \frac{1}{2} \log_2 (2\pi S_1^2) + \log_2 \log_2 + o(1) . \tag{12.8}$$

In the second application of theorem 12.1 we set

$$p_i' = q_{\alpha_L i} \quad (1 \leq i \leq K), \text{ i.e. } \mu_1 = \mathcal{Q}_{\alpha_L}$$
$$a_i = p_i \quad (1 \leq i \leq K), \text{ i.e. } \mu_2 = \mathcal{P}. \tag{12.9}$$

so that M is now $- I(\mathcal{Q}_{\alpha_L} \| \mathcal{P})$. In force of the Neyman-Pearson lemma the infimum γ_L^* (which is now a $\mathcal{P}$-probability) is achieved by the set $\mathcal{B}_L^{(R)c}$, thus coinciding with P_e, i.e. $\log_2 \gamma_L^* = \log P_e$, so that eq. (12.3) gives

$$\log_2 P_e = - L\, I(\mathcal{Q}_{\alpha_L} \| \mathcal{P}) - \frac{1}{2} \log_2 L - \frac{T_2^2}{6 S_2^2} - $$
$$- \frac{1}{2} \log_2 (2\pi S_2^2) + \log \log e + o(1) . \tag{12.10}$$

Remark that if we divide both sides of eq. (12.8) by L we realize that

$$\lim_{L \to \infty} H(\alpha_L) = R = H(\alpha^*)$$

and consequently the sequence $\{\alpha_L\}$ tends to α^* as L tends to the infinity. Moreover from (12.8) we get also (since $R = H(\alpha^*)$):

$$H(\alpha_L) = H(\alpha^*) + \frac{1}{2}\log_2 L + \frac{T_1^3}{6S_1^2} +$$

$$+\frac{1}{2}\log_2(2\pi S_1^2) - \log_2\log_2 e + o(1). \tag{12.11}$$

Expanding $H(\alpha)$ around α^*:

$$H(\alpha) = H(\alpha^*) + H'(\alpha^*)(\alpha - \alpha^*) + o(\alpha - \alpha^*); \tag{12.12}$$

setting α_L instead of α in eq. (12.12) and comparing with eq. (12.11) we get :

$$\alpha_L - \alpha^* = \frac{1}{H'(\alpha^*)}\frac{1}{L}\left\{\frac{1}{2}\log_2 L + \frac{T_1^3}{6S_1^2} +\right.$$

$$\left. +\frac{1}{2}\log_2(2\pi S_1^2) - \log_2\log_2 e + o(1)\right\}. \tag{12.13}$$

On the other hand we can expand also $I(\mathcal{Q}_\alpha \| \mathcal{P})$ around α^*; setting α_L instead of α we have :

$$I(\mathcal{Q}_{\alpha_L}\|\mathcal{P}) = I(\mathcal{Q}_{\alpha^*}\|\mathcal{P}) + (\alpha_L - \alpha^*)\, I'(\mathcal{Q}_{\alpha^*}\|\mathcal{P}) +$$

$$+ o(\alpha_L - \alpha^*). \tag{12.14}$$

Substituting (12.13) into (12.14) for $(\alpha_L - \alpha^*)$ we get :

$$\log_2 P_e = -L\left\{ I(\mathcal{Q}_{\alpha^*}\|\mathcal{P}) + \frac{I'(\mathcal{Q}_{\alpha^*}\ \mathcal{P})}{H'(\alpha^*)}\ \frac{1}{L}\left[\frac{1}{2}\log_2 L + \frac{T_1^3}{6S_1^2} + \right.\right.$$

$$\left.\left. + \frac{1}{2}\log_2(2\pi S_1^2) - \log_2\log_2 e + o(1)\right] + o(\alpha_L - \alpha^*)\right\} - \frac{1}{2}\log_2 L -$$

$$- \frac{T_2^2}{6S_2^2} - \frac{1}{2}\log_2(2\pi S_2^2) + \log_2\log_2 e + o(1).$$

It is only matter of patience to check that

$$I'(\mathcal{Q}_\alpha\|\mathcal{P}) = (\alpha-1)\left[\frac{\sum_1^k{}_i\, p_i^\alpha(\log p_i)^2}{\sum_1^k{}_i\, p_i^\alpha} - \left(\frac{\sum_1^k{}_i\, p_i^\alpha \log p_i}{\sum_1^k{}_i\, p_i^\alpha}\right)^2\right] \qquad (12.16)$$

and since, as we have seen in eq. (10.4)

$$H'(\alpha) = -\alpha\left[\frac{\sum_1^k{}_i\, p_i^\alpha(\log p_i)^2}{\sum_1^k{}_i\, p_i^\alpha} - \left(\frac{\sum_1^k{}_i\, p_i^\alpha \log p_i}{\sum_1^k{}_i\, p_i^\alpha}\right)^2\right]$$

we get

$$\frac{I'(\mathcal{Q}_{\alpha^*}\|\mathcal{P})}{H'(\alpha^*)} = \frac{1}{\alpha^*} - 1\,. \qquad (12.17)$$

Introducing eq. (12.17) into eq. (12.15) and reordering the terms yields eq. (12.5), completing the proof.

Bibliography.

[1] Ariutiunian E.A. Evaluation of Exponent for the Error Probability for a Semi-Continuous Memoryless Channel - Problemi Peredacia Informatzii . 4 (1968) p p. 37-48 (in Russian).

[2] Csiszár J., Longo G. On the Error Exponent for Source Coding etc..., to be published in Studia Math. Acad. Sc. Hung.

[3] Gallager R. Information Theory and Reliable Communication, J. Wiley & Sons, New York, 1968.

[4] Hoeffding W. Asymptotically Optimal Test for Multinomial Distributions. Annals of Math. Stat. 36 (1965), pp. 369-400.

[5] Jelinek F. Probabilistic Information Theory, Mc Graw - Hill, New York, 1968.

[6] Kullback S. Information Theory and Statistics, J. Wiley, New York, 1959.

[7] Rényi A. Wahrscheinlichkeitsrechnung, VEB Deutscher Verlag der Wissenschaften, Berlin 1962.

[8] Straßen V. Asymptotische Abschätzungen in Shannons Informationstheorie, Trans. of the Third

Prague Conference on Information Theory etc..., Prague, 1964, pp. 689-723.

Contents.

Printed in the United States
By Bookmasters